Introductory Soil Science

Laboratory Manual

Dr. Del Dingus

Cal Poly, San Luis Obispo, California

Prentice Hall, Upper Saddle River, NJ 07458

Acquisitions Editor: *Charles Stewart*
Production Supervisor: *Mary Carnis*
Director of Production: *Bruce Johnson*
Production Manager: *Marc Bove*
Production Editor: *Barbara Cassel*

Printed in the United States of America

ISBN 0-13-020080-8

Prentice-Hall International (UK) Limited, *London*
Prentice-Hall of Australia Pty. Limited, *Sydney*
Prentice-Hall Canada Inc., *Toronto*
Prentice-Hall Hispanoamericana, S.A., *Mexico*
Prentice-Hall of India Private Limited, *New Delhi*
Prentice-Hall of Japan, Inc., *Tokyo*
Pearson Education Asia Pte. Ltd., *Singapore*
Editoria Prentice-Hall do Brasil, Ltda., *Rio de Janero*

TABLE OF CONTENTS

ACKNOWLEDGMENT

This laboratory manual is dedicated to Dr. Logan Carter, the founder of the Soil Science Department at Cal Poly, San Luis Obispo in 1949.

Many present and former faculty members have contributed many valuable ideas, and suggestions for developing this manual. Those testing and reviewing this manuscript under class room conditions include Drs. Brent Hallock, Lynn Moody, Thomas Rice, Thomas Ruehr, Terry Smith, and Ron Taskey. Their commitment to quality undergraduate teaching, and high standards for excellence have greatly shaped this manual. Much gratitude is extended to Ms. Joan Stevens for her invaluable assistance with the manuscript. A very humble thank you is extended to present and former students, without whom this laboratory manual would never have been possible.

FOREWORD

Land, air, and water are the vital media supporting humanity. Life exists at an interface among the three. Wherever we journey, whatever we accomplish, we cannot escape our ties to land. The study of soil is the beginning step for understanding our unique environment. It is from plants growing in soil that we obtain three-fourths of our food. Soil supports our buildings and highways; it provides us with medicines to heal our infirmities, bricks to shelter us in comfort, and fiber to cover our imperfections. Soil cleanses both air and water of harmful pollutants and refreshes the land. Soil is a complex habitat for thousands of interesting species, both large and small.

Our goal for this course is twofold: one, to heighten your awareness of the fragile nature of our environment and two, to provide you with the basic technology that will help prepare you for making responsible future decisions in the use of the land, air, and water that sustains us. It is vital, as we manipulate the earth's surface in an effort to improve our quality of life, that each of us share in the responsibility for stewardship of the land. Good land stewardship requires that we seek to understand the natural properties and processes that govern soil. The laboratory exercises that follow are designed to help you learn about these properties and processes. It is only through knowledge and understanding that one can become a responsible steward of our most precious natural resource - *The SOIL*.

Laboratory Organization

Each exercise begins with a goal and a set of objectives. They specify what you should learn or accomplish in the laboratory exercise. To help you gain knowledge about soils, additional information on laboratory topics is **keyed to the 8th edition of the text, Soils in Our Environment, by Miller and Gardiner, Prentice Hall, 1997.** (The designated pages to read are shown in parentheses in each exercise).

The laboratory work will be easier for you if you study the exercise and read the designated pages in the text before coming to class. Plan your work before the session begins.

Make-up labs can be done only during the week a lab is missed. Check with the instructor in charge of the lab you wish to attend for permission to make up a missed lab. Take the quiz, complete the work and have the instructor date and initial your <u>Instructor Copy</u> data sheets.

At the completion of each laboratory period you will be expected to clean and return all equipment to its proper place. You should use the sponge to damp-mop the table top and the surrounding area before leaving. Use dust pans and brushes to pick up spilled soil.

Handle all equipment carefully. Always lay graduated cylinders on their sides and do not invert Erlenmeyer flasks on the table top. No animals, bare feet or food are permitted in the laboratory for reasons of health and safety.

Deionized water is provided at the end of each bench. Only deionized water should be used to fill the plastic wash bottles. The regular tap water from the faucet contains ions which may interfere with the expected results of some of your experiments. Rinse all glassware with deionized water before using.

Quizzes and Laboratory Reports:

Two items will be used to determine your laboratory grade. First, a 10 to 15 minute quiz covering concepts from the previous week's laboratory will be given at the beginning of each session. Second, the instructor copy of pages at the end of each exercise will be collected and evaluated. <u>The quizzes will count as one-half of your laboratory grade.</u> Prompt attendance for the quiz at 10 minutes after the hour is essential. There is no laboratory final for this course.

Your laboratory grade will be reduced to a percent and given to your lecture instructor. The laboratory grade accounts for about 25% of your course grade.

All laboratory reports involving calculations must show solutions to problems sets with complete dimensional analysis. Use extra pages if necessary. Neatness is very important.

Grading:

Weekly Quizzes	50%
Laboratory data sheet evaluations	50%

LABORATORY EXERCISE 1

INTRODUCTION TO SOILS

GOAL: To introduce students to some fundamental physical, chemical, and morphological properties of soils.

Objectives:

1. Know the definitions of all words used in this exercise.

2. Understand and use the Munsell notation for expressing soil color.

3. Know the properties of 6 master horizons of soils (O, A, E, B, C, R) and eight master horizon characteristics designations (p, r, t, k, m, x, g, 2).

4. Examine soil profile features.

5. Understand the relationship between mass, volume, and density, and to determine the density of water and quartz.

6. Understand and be able to calculate the slope steepness expressed in percent, and learn to use a clinometer.

7. Understand the concept of pH, and learn to measure soil reaction using a colorimetric technique.

8. Strengthen your knowledge of the metric system.

9. Solve word problems using percentages, parts per million (ppm), mathematical conversion factors, and dimensional analysis.

LABORATORY EXERCISE I

INTRODUCTION TO SOILS

INTRODUCTION

Soils are three-dimensional bodies of inorganic and organic material generally arranged in layers at the land surface. These layers of topsoil and subsoil, which lie roughly parallel to the ground surface, are called **horizons**. A two-dimensional vertical section of soil extending through the horizons and to the parent material below is called a **profile**. The parent material is the stuff from which the soil forms: it may be solid rock (**residuum**), or unconsolidated material deposited from flowing water (**alluvium**), or from the influence of gravity (**colluvium**), or by moving glaciers (**till**). Earth materials sorted by wind are described as (**eolian**). Sediments accumulated in ancient lakes are known as (**lacustrine**) deposits. Those influenced by ocean processes are described as (**marine**). Organic matter is a source of two other important parent materials for soil: (a.) **peat**, a thick unconsolidated accumulations of fibrous organic matter in lakes, and (b.) **muck**, a lake deposited combination of organic matter mixed with clay. No two soils are exactly alike: they may differ in properties, including color, texture, structure, density, chemistry, depth and horizonation.

(See pages 36-44 and Fig. 2-9 in the Text)

SOIL COLOR

(See pages 90-91 in the Text)

Color is an important physical characteristic of soils. It can be used to differentiate soil horizons, estimate organic matter content, identify soil drainage class, and many other soils conditions. Soil color is identified according to the Munsell Color System as approved by the National Bureau of Standards. The colors are described in three parts: hue, value and chroma, and are always designated in that order.

Hue is the spectral variable. It represents one of the dominant colors of the rainbow, for example, yellow or red. Value represents the relative darkness or lightness of the color. Chroma represents the purity, strength or saturation of a color (colors having zero chromas range from white to gray to black).

Munsell Color Books are used to identify soil color. The book's pages display chips of various colors arranged systematically. Each page represents a different hue, such as 10R, 5YR, 10YR, 2.5YR. The rows on a page designate value. The lower rows (low values) are dark colored and the upper rows (higher values) are light colored. The columns represent the chroma. The left columns (low chroma) are gray colored and the right columns (high chroma) are more brightly colored. Thus, a single hue has many combinations of value and chroma.

The Munsell color notation consists of a color name followed by the sequence hue value/chroma. A brown soil is designated by the notation, brown 10YR 5/3. The name of the color is found on the facing page of the color chart.

Procedure:

1. Obtain a color book and become familiar with the hue, value, chroma, and name designations.

2. Practice by determining the color of the soil samples on the front bench. Compare the color chips in the book with the color of the dry soil first. Then moisten some soil in the palm of your hand and determine its color. Record both the Munsell color notation and color name in Table I below.

Table I. Dry and moist Munsell Color notations and name for three selected soils.

		Sample A	Sample B	Sample C
3.	A. Dry	Notation _____	Notation _____	Notation _____
		Name _____	Name _____	Name _____
4.	A. Moist	Notation _____	Notation _____	Notation _____
		Name _____	Name _____	Name _____

Questions:

1. How does wetting a soil affect a soil color?

2. Which of the three color notations, hue, value, or chroma, is most affected by wetting?

                                                        _____

3. Arrange to following Munsell notations from lightest to darkest.

    10YR7/8, 10YR 2/0, 10YR8/1, 10YR3/6

                          Answer _____

Soil Profiles
(see PP 26-27 in the Text)

A vertical section of soil through all horizons from the surface to the parent material is called a **soil profile**. Horizons are an important distinguishing characteristic among soils. A **horizon** is a layer of soil parallel to the land surface, and different from the layers above and below it. Six major soil horizons are distinguishable by their color, composition and physical characteristics.

O horizons are the layer of organic matter above the mineral soil surface. Many soils have no or only a thin 0 horizon either because organic material is rapidly decomposed, or very little is deposited on the soil surface. Forest soils may have an appreciable accumulation of organic matter (0 horizon).

A **horizons** are mineral horizons characterized by a dark color caused by an accumulation of decomposed organic matter (**humus**) mixed in the mineral soil. Under favorable conditions "A" horizons are very biologically activity.

E **horizons** are lighter in color than the A above or B below. They are zones of **eluviation**, that is, fine clay and organic material have been transported by percolating water and deposited in a lower horizon. The E horizon is unusual and often found in sandy soils in cold climates with high rainfall and conifer forests.

B **horizons** are mineral horizons which usually occur beneath an A or E horizon and above a C. They are zones of **illuviation**, an accumulation of clay, iron, aluminum or humus, or combinations of these materials. Some soils lack a B horizon, because they are so young that illuvial materials have not had time to accumulate.

C **horizons** are the layers of unconsolidated earth material, either unaltered transported material(alluvium, colluvium, dune sand, loess, till...) or bedrock which has weathered in place.

R **horizons** are the underlying hard unweathered bedrock material. Granite, basalt, sandstone, and limestone are examples.

The major horizons are often subdivided to note special soil features, for example: Ap, Bt, Cr, and Bk. The letter "p" is associated with the A horizon and indicates a mixed horizon. The letter "t" is associated with the B horizon and signifies an accumulation of clay. The "k" designates an accumulation of lime in the B or C horizon and the letter "r", associated with the C, indicates weathered rock.

Note: The soil horizon designations were revised in 1982. Books and soil surveys published before 1982 used the old system. We will emphasize the new system here. Refer to the textbook for a more complete list of soil horizon symbols and their meanings.
(see pages 44-45 in the Text)

Procedure:

Examine two soil **monoliths** (profiles) assigned by the instructor. Record your observations of the horizon designations and their depth in centimeters from the soil surface. The dry soil color (Munsell notation), presence of rock fragments, cracks (notice any pattern), and roots should be included in a sketch of the profile.

Soil Profile 1. Soil Series Name _____

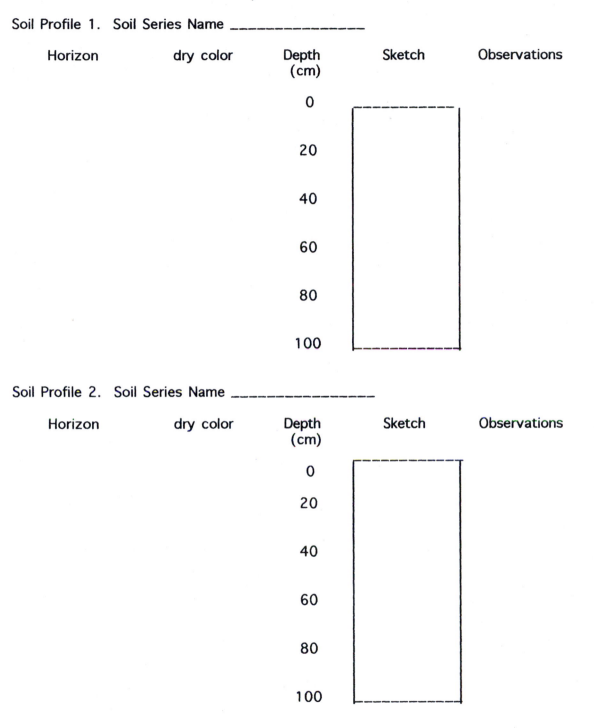

Horizon	dry color	Depth (cm)	Sketch	Observations
		0		
		20		
		40		
		60		
		80		
		100		

Soil Profile 2. Soil Series Name _____

Horizon	dry color	Depth (cm)	Sketch	Observations
		0		
		20		
		40		
		60		
		80		
		100		

Questions:

1. What horizon characteristic is designated by these symbols: p, t, k, r, m, x, s, and g?

   p _____   m _____

   t _____   x _____

   k _____   s _____

   r _____   g _____

2. Place the letter of the master horizons usually associated with p, r, k, x, 2 and t designations. (example Bx,   a soil horizon having a brittle hardpan), See Text Table 2-2, PP 45.

   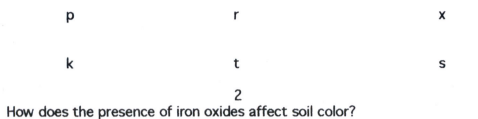

       p   r   x

       k   t   s

                            2
3. How does the presence of iron oxides affect soil color?

4. Which horizon (A, E or B) is usually first to form from a parent material as the soil develops?

Density

Density is an important physical property of soils. Compacted soils have greater density than non-compacted soils. Compacted soils inhibit root penetration, decrease water movement through the soil, and are difficult to cultivate. Also, density is useful for comparing soils on a weight (mass) and volume basis. **Density** is defined as the mass of a substance divided by its volume.

$$\text{Density} = \frac{\text{Mass}}{\text{Volume}}$$

Water has arbitrarily been chosen as the standard for comparison; its density is 1.0 gram per cubic centimeter (1.0 g/cm^3). For convenience, one milliliter of liquid water at 25°C equals 1 cubic centimeter of volume.

Table 2. Densities of some common materials

```
_____
        Material           Density (g/cm³)
_____

        Wood                   0.8
        Ice                    0.9
        Soil                   1.3
        Sandstone              2.6
        Aluminum               2.7
        Steel                  7.9
        Lead                  11.3
        Mercury               13.4
        Gold                  19.2

_____
```

Procedure to measure the density of water:

1. Place between 92 and 98 ml of tap water into a 100 ml graduated cylinder.

2. Tare a balance to zero.

3. Place the graduated cylinder on the balance and determine the mass of the cylinder plus water to the nearest tenth (0.1) of a gram. Record the mass and water volume (to the nearest 0.5 ml) on the following data page.
    Note: the volume of a liquid is taken at the bottom of the meniscus. The meniscus is the curved surface of the liquid.

4. Pour out most, but not all of the water. Leave about 10 to 15 ml of water in the cylinder. Record the new volume and mass.

TABLE FOR RECORDING THE DENSITY OF WATER

		initially full	partially empty
a.	Mass of cylinder + water	_____ g	_____ g
b.	Volume of water	_____ ml	_____ ml
c.	Net mass of water (difference between initially full and partially empty)	_____ g	
d.	Net volume of water (difference between initially full and partially empty)	_____ ml	

e. Calculate the density of water to two decimal places using your data.

Calculate density of water   Answer   _____ g/cm³
(neatly show all steps and units)

Procedure to measure the density of quartz:
1. Fill a 100 ml graduated cylinder approximately half full with water. Record below the exact volume of water in the cylinder to the nearest 0.5 mls.

2. Weigh the cylinder and water. Record the mass to the nearest 0.1 gram.

3. Add three pieces of quartz to the cylinder of water. Record the new volume and the new mass.

4. Determine the volume and mass of the quartz.

5. Calculate the density of quartz to two decimal places. Include the proper units, and show the complete set-up for the calculation.

TABLE FOR RECORDING THE DENSITY OF QUARTZ

Volume: water only   _____ml

                  water plus quartz   _____ml

                  quartz   _____cm^3

Mass: Cylinder plus water plus quartz   _____g

                  cylinder plus water   _____g

                  quartz only   _____g

Density of quartz   _____g/cm^3
(neatly show calculations)

Questions:

1. Which can be measured more precisely with the equipment used in this laboratory, mass or volume? Why?

2. The density of water is 1.0 g/cm^3. Convert this density into the English units of pounds per cubic foot for water: (memorize this conversion and value).

3. How many times greater is the density of quartz than water?   _____

4. A 358 gram sample of a gray metallic material displaced 31.68 cm^3 of water. From Table 2 above it is most likely composed of what material?

                                                   _____

Slope Measurements

Steepness of slope is a major component of a landscape's topography. It influences the rate of soil development and the potential for soil erosion. Slope expresses a change in vertical height over a horizontal distance, and is recorded in percent or in degrees, without a (+) or (-) designation. Soil scientist usually express slope in percent as shown below.

Rise = 3 units   Run = 4 units

Slope = Rise/Run   Note: 3/4 = 75/100 = 75%
                           or 3/4 = 0.75 X 100 = 75%

The slope of a landscape can be measured with a clinometer. To use the clinometer, sight up slope or down slope to an object that is at the same height above ground as your eye. The distance should be between 10 and 20 meters away, although the exact distance is not critical. Read the % slope on the right scale. The left scale expresses slope in degrees.

Procedure:

Practice using the clinometer by sighting on the following objects chosen by the instructor:

Object	Reading (% slope)
A. _____	_____
B. _____	_____
C. _____	_____
D. _____	_____

Questions:

1. Calculate the steepness of a slope when the rise vertically is 32 meters over a horizontal distance of 308 meters.

2. A 100% slope is _____ degrees.

3. The horizontal distance along the base of a symmetrical hill is 1400 meters. The slope up the hill from the base to the top is 12 percent. Draw its picture and find the height of the hill.

_____ m

Soil pH
(See pages 168-171 in Text)

The pH, or soil reaction, is probably the most important and easily measured chemical property of soil. It is used mainly to evaluate the suitability of soil for crop growth. The pH value of a soil may suggest to the grower that more detailed chemical testing for plant available nutrients and management problems will be needed.

The pH scale, ranging from 0 and 14, is a measure of the hydrogen ion (H^+) activity in the soil solution. The pH is defined as the negative logarithm of the (H^+) activity ($-\log (H^+)$), where the hydrogen ion activity is measured in units of grams/liter. The logarithmic function means that for each unit change in pH, the H^+ activity changes by a factor of 10. If a solution contains 1×10^{-7} grams H^+/liter, its pH is 7 (the negative of the concentration's logarithm, - one times a -7). Acids are solutions that contain more H^+ ions than OH^- (hydroxide) ions.

pH scale

```
0   1   2   3   4   5   6   7   8   9   10  11  12  13  14
_____acidic_____neutral_____basic or alkaline_____
```

```
        |_____|
         pH range of most soils
```

Measuring soil pH with the Poly-D pH kit

Procedure:

1. Use a spatula, to place a small amount of crushed soil into the depression of a spot plate found with the pH test kit.

2. Add a few drops of the Poly-D pH indicator to flood the soil. Tilt the spot plate back and forth for about 30 seconds.

3. Match the color of the liquid to the closest color on the pH color chart and record.

   Soil A _____   Soil B _____   Soil C _____

Questions on pH:

1. What is the optimal pH range for the following plants: (see the chart)

   Plant: _____   _____   _____   _____

   Optimum pH range: _____   _____   _____   _____

2. Which is most acidic, pH 4, pH 6, pH 7.7, pH 9.1? (circle answer)

3. What compound is formed when H^+ combines with OH^-? _____

4. What is the direction and magnitude of change in hydrogen ion concentration when a soil at pH of 5.0 is raised to pH 8.0 by the addition of lime?

   The H^+ concentration (___ increases, ___ decreases) by a factor of _____ .

Dimensional Analysis

With the existence of many different systems of weights and measures it becomes important for an individual to learn how to convert from one system to another. The American Society of Agronomy uses the International System of Units. Other scientific and industrial organizations have adopted metric units. The United States is one of the few countries still using the British System. Most countries are now using the metric system to conduct international trade. The Metric Conversion Act of 1975, international trade, and scientific communication necessitate knowing how to convert from one system to another.

Conversions can be done logically if one uses the process of dimensional analysis. This process is quite simple and consists of assembling all of the facts, putting them into a logical sequence and then performing the mathematical operations in a continuous process. It is based upon the concept of proportions. A fundamental principle of dimensional analysis is that $A/A = 1$ and that $B \times 1 = B$.

Consider this example involving time and distance.

How long in hours would it take to send a message from here to a space ship near Pluto? Pluto is about three billion miles from Earth and message signals travel at 1.86×10^5 miles per second.

$$3 \times 10^9 \text{ miles} \times \frac{1 \text{ second}}{1.86 \times 10^5 \text{ miles}} \times \frac{1 \text{ minute}}{60 \text{ seconds}} \times \frac{1 \text{ hour}}{60 \text{ minutes}} = 4.5 \text{ hours}$$

(notice all the units cancel except hours)

Percentages and parts per million (ppm):

Soil data are often reported as percent (%) or as parts per million (ppm). Percent is the same as parts per hundred. Very low concentrations of material in soil are reported in ppm. Four ppm boron in a soil means that 1 million kilograms of the soil contains 4 kilograms of boron. An hectare-furrow-slice of soil weighs 2 million kilograms, 4 ppm boron in the soil equals 8 kilograms of boron per acre-furrow-slice.

CONVERSIONS USEFUL FOR INTRODUCTORY SOIL SCIENCE

Both metric and the British system of weights and measures are used to express quantities in soil science. Eventually only one system will be in use. For now, it is advantageous to learn both systems during the transition period.

Two basic rules apply to the use of the metric system of weights and measures. First, all units of measurement in the metric system are based upon the number 10 and its multiples. Second, orders of magnitude for numbers are often expressed as powers of 10. The use of prefixes eliminates the need to write numbers in scientific notation. For example, the number 0.0041 is written as 4.1×10^{-3} in scientific notation. The use of metric prefixes allows 0.0041 grams to be written as 4.1 milligrams. Question: How many milligrams are in one gram?

Metric units vary from each other by factors of 10, i.e., 10 millimeters = 1 centimeter; 100 centimeters = 1 meter; 1000 grams = 1 kilogram. The following prefixes are used to establish the magnitude of a number in relation to 1.0.

TABLE 3. Prefixes, decimal equivalents and symbols used with the metric system.

Prefix	Decimal Equivalent	Scientific Notation	Symbol
micro	0.000001	1×10^{-6}	μ
milli	0.001	1×10^{-3}	m
centi	0.01	1×10^{-2}	c
deci	0.1	1×10^{-1}	d
deka	10.0	1×10^{1}	da
hecto	100.0	1×10^{2}	h
kilo	1000.0	1×10^{3}	k
mega	1000000.0	1×10^{6}	M

Examples of calculations using scientific notation.

Addition	$(2 \times 10^{6}) + (6 \times 10^{6}) = 8 \times 10^{6}$
Subtraction	$(8 \times 10^{4}) - (4 \times 10^{4}) = 4 \times 10^{4}$
Multiplication	$(5 \times 10^{3}) \times (6 \times 10^{4}) = 3 \times 10^{8}$
Division	$(9 \times 10^{6}) / (3 \times 10^{2}) = 3 \times 10^{4}$

TABLE 4: Common equivalent British and metrics units used in agriculture.

Length		Area	
1 micrometer	$= 1 \times 10^{-6}$ meters	1 section	$=$ 640 acres
1 centimeter	= 0.39 inches	1 hectare	$=$ 10,000 m^2
1 centimeter	= 10 mm	1 hectare	$=$ 2.47 acres
1 meter	= 1.094 yards	1 acre	$=$ 43,560ft^2
1 meter	= 100 centimeters		
1 kilometer	= 1000 m	Temperature	
1 kilometer	= 3280.8 feet	$°F = 1.8 °C + 32$	
1 inch	= 2.54 cm	$°C = (°F - 32)/1.8$	
1 foot	= 30.48 cm		
1 mile	= 5280 feet		

mass

1 kilogram	= 1000 g	1 pound	= 453.6 grams
1 kilogram	= 2.205 pounds	1 pound	= 16 ounces
1 metric tonne	= 1000 kg	1 ounce	= 28.3 grams
1 U.S. ton	= 2000 pounds	1 cwt.	= 100 pounds
1 quintal	= 100 kilograms		

Volume		Density of Water and Soil
1 ml	= 1 cm^3	1 ml of water = 1 gram
1 cubic meter	$= 1 \times 10^6$ cm^3	1 liter water = 1 kilogram
1 liter	= 0.265 gallon	1 cubic ft. water = 62.4 pounds
1 liter	= 0.945 quart	*1 hectare-15 cm $= 2 \times 10^6$ kg
1 liter	= 1000 cm^3	*1 acre furrow slice $= 2 \times 10^6$ lb
1 cubic foot	= 28.305 liters	1 acre-foot = 43,560 cubic feet
1 cubic meter	= 1000 liters	

*Based upon assumed average bulk density values

Definitions:

Density   mass/volume (g/cm^3, lbs/ft^3, kg/m^3 etc.)

Percent   The ratio of amounts expressed as parts per 100

example $5/20 = X/100$, $X = 25$

$120/560 = X/100$ $X = 21.4$

PPM   One part of a substance per million parts total

examples: 1 milligram/kilogram
1 milligram/liter water
1 microgram/gram
1 microgram/milliliter water
1 pound/million pounds
1 gallon/million gallons

Note: In every case it takes one million units of the numerator to equal one unit of denominator.

To convert 5000 ppm to pph use ratio and proportions

Examples:

a. $5000/1,000,000 = X/100$ $X = .005$ or 0.5%

b. $X/1,000,000 = 3/100$ $X = 30,000$

LAB PROBLEM SET I

INSTRUCTIONS: Do each of these problems on notebook paper. Show the complete setup using dimensional analysis whenever appropriate. Draw a box around your answer. <u>Turn in these pages with the answers in the blanks and attach the written work to the back.</u>

1. A. $(3.5 \times 10^6)(1.2 \times 10^2)$ = _____

    B. $(4.1 \times 10^7)(8.6 \times 10^{-3})$ = _____

    C. $(2.8 \times 10^{-4})(6.5 \times 10^{-2})$ = _____

    D. $[(3.5 \times 10^6)/(2.7 \times 10^{-3})] \times [1.4 \times 10^{-6}]$ = _____

    E. $[(2.4 \times 10^4) + (8.3 \times 10^4)]/[(6.7 \times 10^2) - (1.2 \times 10^1)(2.5 \times 10^1)]$

    = _____

2. Feet (British system) and meters (metric system) are the conventional units used to express length. The dimensions of a lab are 46 feet long by 28 feet wide. What are its dimensions in the metric system?

    a) _____ m. long _____ by m. wide

    b) _____ cm. long _____ by cm. wide

    c) What is your height in inches, in meters, in centimeters?

    _____ in _____ cm _____ m

3. The common unit of measurement for the area of agricultural land is the **acre**. Since most fields contain many acres it's difficult to visualize the size of one acre. Most textbooks now use **hectares**. To better understand the relationship of acres and hectares calculate (a) and (b) below. Draw a diagram of a football field and a baseball field.

    a) A regulation football field is 100 yards from goal line to goal line and 160 feet wide. Calculate the area of the football field in:

    square feet = _____ , acres = _____,

    square meters = _____, hectares = _____

15

b). The approximate size of a regular baseball park can be obtained by considering the playing field as a quarter of a circle centered at home plate. The outfield fence would average about 373 feet from home plate. The area of a full circle $= \pi r^2$ where $\pi = 3.14$ and r = the radius. Calculate the area of the baseball field in:

    square feet = _____,   acres = _____,

    square meters = _____,   hectares = _____

4. A section of land is equal to one square mile. Determine the number of acres in one section of land.
        a section equals _____ acres.

5. An average yield reported for alfalfa is 5.0 tons (U.S.) per acre. Calculate the equivalent yield in terms of the following units.

    a.) kilograms/hectare (kg/ha)   = _____

    b.) metric tonnes/hectare (MT/ha) = _____

Observe: Metric tonnes/ha is approximately twice as large as tons (U.S.)/ac. Kilograms/ha is approximately equal (only slightly larger) than pounds/ac.

6. Soil temperatures commonly range from freezing to very warm depending upon the vegetative cover, soil moisture content and the amount of sunshine reaching the soil surface. Calculate the following soil temperatures.

    a.) minimum for the year 20° F   = _____ °C

    b.) maximum for the year 110° F   = _____ °C

    c.) change 24.6 °C to   = _____ °F

7. Calculate the equivalent depth of water in mm for the reported precipitation at San Luis Obispo, California.

	inches	millimeters
Driest year 1898	_____	176.02 mm
Wettest year 1968/69	54.53	_____
Average annual precipitation	_____	523.94 mm

8. a. Convert 2.8 % to ppm _____

    b. Convert 15,000 ppm to parts per 100 _____

Instructor Copy

Name _____

Section _____

Table I. Dry and moist Munsell Color notations and name for three selected soils.

	Sample A	Sample B	Sample C

3. A. Dry   Notation _____   Notation_____   Notation_____

Name _____   Name _____   Name_____

4. A. Moist   Notation _____   Notation_____   Notation_____

Name _____   Name _____   Name_____

1. How does wetting a soil affect a soil color?

2. Which of the three color notations, hue, value, or chroma, is most affected by wetting?

_____

3. Arrange to folowing Munsell notations from lightest to darkest.

10YR7/8, 10YR 2/0, 10YR8/1, 10YR3/6

Answer _____

Soil Profile 1:   Soil Series Name _____

Horizon   dry color   Depth   Sketch   Observations
                                          (cm)

Laboratory I

Instructor Copy

Name _____

Section_____

Soil Profile 2:   Soil Series Name _____

Horizon	dry color	Depth (cm)	Sketch	Observations

1. What horizon characteristic is designated by p, t, k, r, m, x, s, and g?

    p _____   m _____

    t _____   x _____

    k _____   s _____

    r _____   g _____

2. Place the letter of the master horizons usually associated with p, r, k, x, s and t designations. (example Bx,   a soil horizon with having a brittle hardpan), See Text Table 2-2, PP 45.

    p   r   x

    k   t   s

3. How does the presence of iron oxides affect soil color?

4. Which horizon (A, E or B) is usually first to form from a parent material as the soil develops?

Laboratory I

Instructor Copy

Name _____

Section_____

TABLE FOR RECORDING THE DENSITY OF WATER

		initially full	partially empty
a.	Mass of cylinder + water	_____ g	_____ g
b.	Volume of water	_____ ml	_____ ml
c.	Net mass of water (difference between initially full and partially empty)	_____ g	
d.	Net volume of water (difference between initially full and partially empty)	_____ ml	

e. Calculate the density of water to two decimal places using your data.

Calculate density of water   Answer   _____ g/cm^3
(neatly show all steps and units)

TABLE FOR RECORDING THE DENSITY OF QUARTZ

Volume:	water plus quartz	_____ ml
	water only	_____ ml
	quartz	_____ cm^3
Mass:	Cylinder plus water plus quartz	_____ g
	cylinder plus water	_____ g
	quartz only	_____ g
Density of quartz (neatly show calculations)		_____ g/cm^3

Laboratory 1

Instructor Copy

Name _____

Section_____

Questions:

1.   Which can be measured more precisely with the equipment used in this laboratory, mass or volume? Why?

2.   The density of water is 1.0 g/cm^3. Convert this density into the English units of pounds per cubic foot for water: (memorize this conversion and value).

3.   How many times greater is the density of quartz than water?   _____

Clinometer Readings

Object   Reading (% slope)

A. _____ _____

B. _____ _____

C. _____ _____

D. _____ _____

Questions:

1.   Calculate the steepness of a slope when the rise vertically is 32 meters over a horizontal distance of 308 meters.

                                                            _____

2.   A 100% slope is _____ degrees.

3.   The horizontal distance along the base of a symmetrical hill is 1400 meters. The slope up the hill from the base to the top is 12 percent. Draw its picture and find the height of the hill.

                                                    _____ m

Instructor Copy   Name _____

                                         Section_____

3.   Match the color of the liquid to the closest color on the pH color chart and record.

        Soil A _____   Soil B _____   Soil C _____

Questions on pH:

1.   What is the optimal pH range for the following plants: (see the chart)

     Plant: _____   _____   _____   _____

     Optimum pH range: _____   _____   _____   _____

2.   Which is most acidic, pH 4, pH 6, pH 7.7, pH 9.1? (circle answer)

3.   What compound is formed when H^+ combines with OH^- ? _____

4.   What is the direction and magnitude of change in hydrogen ion concentration when a soil at pH of 5.0 is raised to pH 8.0 by the addition of lime?

     The H^+ concentration (____ increases, ____ decreases) by a factor of _____ .

LAB PROBLEM SET I

Instructor Copy

Name _____

Section_____

INSTRUCTIONS: Do each of these problems on notebook paper. Show the complete setup using dimensional analysis whenever appropriate. Draw a box around your answer. Turn in these pages with the answers in the blanks and attach the written work to the back.

1. A. $(3.5 \times 10^6)(1.2 \times 10^2)$ = _____

    B. $(4.1 \times 10^7)(8.6 \times 10^{-3})$ = _____

    C. $(2.8 \times 10^{-4})(6.5 \times 10^{-2})$ = _____

    D. $[(3.5 \times 10^6)/(2.7 \times 10^{-3})] \times [1.4 \times 10^{-6}]$ = _____

    E. $[(2.4 \times 10^4) + (8.3 \times 10^4)]/[(6.7 \times 10^2) - (1.2 \times 10^1)(2.5 \times 10^1)]$

    = _____

2. Feet (British system) and meters (metric system) are the conventional units used to express length. If the dimensions of a lab were 46 feet long by 28 feet wide. What are its dimensions in the metric system?

    a) _____ m. long _____ by m. wide

    b) _____ cm. long _____ by cm. wide

    c) What is your height in inches, in meters, in centimeters?

    _____ in _____ cm _____ m

3. The common unit of measurement for the area of agricultural land is the **acre**. Since most fields contain many acres it's difficult to visualize the size of one acre. Most textbooks now use **hectares**. To better understand the relationship of acres and hectares calculate (a) and (b) below. Draw a diagram of a football field and a baseball field.

    a)   A regulation football field is 100 yards from goal line to goal line and 160 feet wide. Calculate the area of the football field in:

    square feet   = _____   acres   = _____

    square meters = _____   hectares   = _____

b). The approximate size of a regular baseball park can be obtained by considering the playing field as a quarter of a circle centered at home plate. The outfield fence would average about 373 feet from home plate. The area of a full circle = πr^2 where π = 3.14 and r = the radius. Calculate the area of the baseball field in:

square feet = _____   acres = _____

square meters = _____   hectares = _____

4. A section of land is equal to one square mile. Determine the number of acres in one section of land.

a section equals _____ acres.

5. An average yield reported for alfalfa is 5.0 tons (U.S.) per acre. Calculate the equivalent yield in terms of the following units.

a.) kilograms/hectare (kg/ha) = _____

b.) metric tonnes/hectare (MT/ha) = _____

Observe: Metric tonnes/ha is approximately twice as large as tons (U.S.)/ac. Kilograms/ha is approximately equal (only slightly larger) than Lbs/ac.

6. Soil temperatures commonly range from freezing to very warm depending upon the vegetative cover, soil moisture content and the amount of sunshine reaching the soil surface. Calculate the following soil temperatures in degrees Celsius.

a.) minimum for the year 20° F = _____ °C

b.) maximum for the year 110° F = _____ °C

c.) change 24.6 °C to = _____ °F

7. Calculate the equivalent depth of water in mm for the reported precipitation at San Luis Obispo, California.

	inches	millimeters
Driest year 1898	_____	176.02 mm
Wettest year 1968/69	54.53	_____
Average annual precipitation	_____	523.94 mm

8. a. Convert 2.8 % to ppm _____

b. Convert 15,000 ppm to parts per 100 _____

LABORATORY EXERCISE II

MINERALS, ROCKS AND WEATHERING

Goal: To familiarize students with important soil-forming minerals and rocks, and the weathering processes.

Objectives:

1. Understand the definition of all words used in this exercise.

2. Learn the chemical and physical properties of common minerals and the major rock types important in soil formation.

3. Identify each of the following:

quartz	biotite	hornblende	sandstone	granite	schist
feldspar	muscovite	calcite	limestone	basalt	serpentinite

4. Know the 26 elements required by plants and animals.

5. Learn:
    2 minerals that supply K   a mineral that supplies P
    2 minerals that supply Fe   a mineral that supplies S
    2 minerals that supply Ca   a mineral that supplies Mg

6. Explain how and why most soils gradually become more acidic with time.

7. Know chemical reaction for the formation of carbonic acid.

8. Know the influence of oxidation and reduction on soil color.

9. Examine the changes that occur when rocks weather to form parent materials that eventually develop into soil.

10. Know the relationship between soil drainage and the oxidation-reduction of iron.

MINERALS, ROCKS AND THEIR WEATHERING

Most soils form by the physical **disintegration** and chemical **decomposition** of minerals and rocks (the exception is that group of soils which form almost entirely from organic matter). These physical and chemical processes are called weathering. Thus, minerals and rocks may serve as soil parent materials by weathering directly in place to form soil. More commonly, they break down physically, and then are transported to a new location where soil is formed by further physical and chemical weathering.

Rocks are made up of minerals, and all minerals are composed of one or more elements. Many of these elements are essential for plant and animal growth. The natural fertility of soils depends on the release of these elements by weathering. Although minerals contain all 26 life essential elements, only 22 are made available by weathering of minerals and rocks.

The elements required by plants and the additional elements required by animals are listed in Table 5. The chemical symbol of each essential element is given along with the ionic form utilized by plant. Plants contain many more elements than they need. Animals obtain most of their nutrients either directly from plants or by consuming other herbivores. They require all the plant essential elements, except boron, plus 10 more.

(See Section 10-2 page 312-313 in the Text)

Table 5. ELEMENTS ESSENTIAL FOR PLANTS AND ANIMALS

ESSENTIAL FOR PLANTS AND ANIMALS			ESSENTIAL FOR ANIMALS	
SYMBOL	ELEMENT	FORM	SYMBOL	ELEMENT
C	Carbon	CO_2	Na	Sodium
H	Hydrogen	H , H_2O	Si	Silicon
O	Oxygen	H_2O, O_2	As	Arsenic
N	Nitrogen	NO_3^- or NH_4^+	F	Fluorine
P	Phosphorus	$H_2PO_4^-$ HPO_4^{--}	I	Iodine
K	Potassium	K^+	Co	Cobalt
Ca	Calcium	Ca^{++}	Ni	Nickel
Mg	Magnesium	Mg^{++}	Se	Selenium
S	Sulfur	SO_4^{--}	Cr	Chromium
Fe	Iron	Fe^{+++}	Sn	Tin
Mn	Manganese	Mn^{++}		
B	Boron	$B(OH)_3$*		
Mo	Molybdenum	MoO_4^{--}		
Cu	Copper	Cu^{++}		
Zn	Zinc	Zn^{++}		
Cl	Chlorine	Cl^-		

----

* Boron has not been proven essential for animals.

NOTE: (+) charged ions are called **cations** and (-) charged ions are called **anions**

Plants also absorb variable amounts of other elements that have not been proven essential for life. Some can be toxic (i.e., Al, aluminum, Cs, cesium) others can be **beneficial**(i.e. V, Vanadium, Co, Cobalt, Ni, Nickel). An essential element is needed by an organism for normal growth and reproduction. A beneficial element, though not required, can enhance the growth and performance of living organisms.

(See table 10-1 in the Text)

Two useful mnemonics:
C HOPKNS CaFe Mg B Mn CuZn Mo's Cl
I Cr F Si Se As NiCo Sn Na (I Cram For Science Seminar As NiCol Sneakes a Nap)

PART 1. MINERAL EXAMINATION

A mineral is a naturally occurring, crystalline, inorganic solid having a definite chemical composition and specific physical properties. Minerals can be identified by studying their physical properties including hardness, color, and cleavage, the tendency to split along certain predictable directions. Once the name of a mineral is identified by examining these and other physical properties, its chemical composition can be determined from a table of minerals. The relative ease of weathering of a mineral can be predicted from a knowledge of its properties, most importantly hardness and chemical composition.

Plants rarely obtain the elements carbon, hydrogen, oxygen and nitrogen from minerals. Instead, plants obtain C from carbon dioxide (CO_2) in the atmosphere, H from water (HOH), O from oxygen gas (O_2) in the atmosphere or from water, and N from organic matter decomposition or from air via a nitrogen fixing process performed by a few microorganisms. Plants cannot use atmospheric N_2 nitrogen directly. Although C, H, O, N are found in many minerals they are not made available to plants by weathering.

Procedure:

Complete **Part 1, Mineral Examination,** by studying the mineral specimens. Use Table 6 for determining the plant nutrient elements contained in each mineral. In the column at the right, place the chemical symbol for each nutrient made available to plants by the weathering of the mineral.

Use the following guide for **hardness:**

A hard mineral cannot be scratched by a knife, it will scratch glass.

A medium mineral can be scratched by a fingernail file.

A soft mineral can be scratched by a penny.

MINERAL EXAMINATION

Mineral	Hardness hard, med, soft	Color	Cleavage yes or no	Plant nutrients released by weathering
Quartz				
Orthoclase				
Plagioclase				
Muscovite				
Biotite				
Hornblende				
Hematite				
Pyrite				
Apatite				
Calcite				
Gypsum				

Mineral Questions:

1. Which mineral is most resistant to weathering? _____

2. Name two feldspar minerals. _____ _____

3. Identify two micas. _____ _____

4. List two ferromagnesian minerals _____ _____

Table 6. Common Soil Minerals and their Chemical Composition

MINERAL	CHEMICAL COMPOSITION	SPECIMEN NO.
Quartz (Milky)	SiO_2	1-4
Feldspars		
a. Orthoclase	$KAlSi_3O_8$	11
b. Plagioclase	$(Na, Ca)AlSi_3O_8$	12
Micas		
a. Muscovite	$KAl_2(Al,Si)_3O_{10}(OH)_2$	7
b. Biotite	$K(Mg.Fe)_3(Al,Fe)Si_3O_{10}(OH)_2$	8
Ferro-Magnesian		
a. Hornblende (an amphibole)	$(Ca,Na)(Mg,Fe)_3(Si,Al)_4O_{11}(OH)_2$	13
b. Augite (a pyroxene)	$(Ca,Na)(Fe,Mg)(Al,Si)_2O_6$	14
Pyrite	FeS_2	20
Apatite	$Ca_{10}(PO_4)_6(F,Cl,OH)_2$	9
Iron Oxides		
a. Hematite	Fe_2O_3	18
b. Limonite	$2Fe_2O_3*3H_2O$	17
c. Magnetite	Fe_3O_4	19
Talc	$Mg_3Si_4O_{10}(OH)_2$	6
Serpentine	$Mg_3Si_2O_5(OH)_4$	5
Chalcopyrite	$CuFeS_2$	21
Sphalerite	ZnS	22
Calcite	$CaCO_3$	15
Dolomite	$CaMg(CO_3)_2$	16
Gypsum	$CaSO_4 * 2H_2O$	24

Note: The other essential elements for plants and animals not found here occur in trace amounts in these and other minerals.

PART II. ROCK EXAMINATION

Most **rocks** are assemblages of two or more minerals. They are classified on the basis of their origin into three categories: (A) **igneous**, (B) **sedimentary** and (C) **metamorphic**. Igneous rocks are further subdivided into visual texture classes based upon the size of minerals in the rock. Coarse grained rocks are composed of minerals distinguishable with the unaided eye. Fine grained rocks contain minerals that are recognizable only by magnification. A description and examples of each are given below. This lab emphasizes sedimentary and igneous rocks because they are the most abundant source of soil parent materials.

A. Igneous Rocks

Igneous rocks are formed by the cooling and solidification of **magma**, a molten rock material originating deep in the earth. Igneous **intrusive** rocks form by the slow cooling of magma deep in the earth. The slow cooling allows mineral crystals to grow fairly large- at least large enough to be recognized with the naked eye; hence, intrusive rocks are coarse grained. Igneous **extrusive** rocks form by rapid cooling of magma (lava) at or near the surface of the earth. The rapid cooling arrests mineral growth before crystals grow large enough to be recognized with the naked eye. Extrusive rocks are very fine grained. Igneous rocks are further classified on the basis of a magma's chemistry and the mineralogy of the resultant rock. Two broad distinctions are **felsic** and **mafic** igneous rocks. Felsic rocks are higher in silica and light colored. Mafic rocks are lower in silica and dark colored. For soil formation the two most important igneous rocks, in term of abundance, are granite and basalt.

A simplified classification scheme for crystalline Igneous rocks is given below:

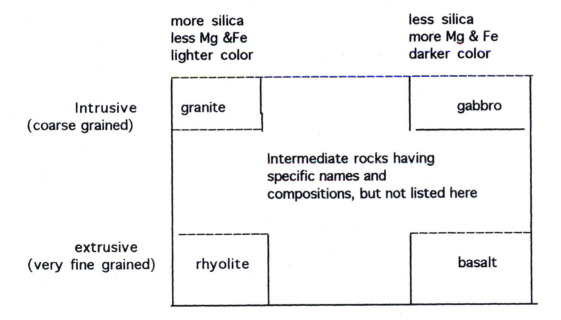

(See pages 1-7 and Figures 1.1 and 1.2 in the Text)

Granite is a common coarse grained (large crystal) igneous intrusive rock. It is composed largely of quartz and feldspars (orthoclase and plagioclase) with small amounts of accessory minerals, including muscovite and biotite. The presence of dark biotite among the light colored minerals often gives granite a characteristic speckled appearance. Granite often weathers into coarse, sandy soils high in quartz and feldspar.

Basalt is the most prevalent fine-grained (tiny crystal) igneous extrusive rock. It is a common volcanic rock. Minerals in basalt are microscopic and consist mostly of plagioclase and augite often with small amounts of magnetite and pyrite. Basalt often breaks down into reddish colored soils having high clay content. The abundance of small mineral crystals and the high iron content of basalt causes it to weather rapidly.

B. Sedimentary Rocks

Sedimentary rocks are those that form by one of two processes: 1) erosion, transport, deposition and burial lithification of fragments of igneous, metamorphic or other sedimentary rocks; or 2) direct precipitation and cementation of biological or chemical materials. The first group includes conglomerates, sandstones and shales cemented together by calcium carbonate, silica, or iron oxide. The second group includes limestone and travertine, formed from biological or chemical accumulations of calcium carbonate. Sedimentary rocks are the most abundant material on the earth's land surface; hence, more soils form from sedimentary rocks than from any other type.

1. **Conglomerates** consist of rounded pebbles (> 2 mm in diameter) or cobbles mixed with sand, silt, and clay. They are cemented together by fine grained silica, carbonates or oxides. Conglomerates weather to gravely, coarse textured soils high in quartz.

2. **Sandstones** consist of grains of sand- sized (2 mm to 0.05 mm in diameter) quartz, feldspars, and sometimes micas, mixed with some silt and clay, and bound together by various cementing agents. Sandstones may weather to either coarse or fine textured soils, depending on the minerals and amount of silt and clay in the rock. Many soils in California that have formed from sandstone are fine textured.

3. **Shales** consist of consolidated silt and clay; hence, soils derived from shales are either silty or clayey, and usually contain significant amounts of the clay minerals montmorillonite and kaolinite. Although shales are often soft, they may be hard if cemented by silica.

4. **Limestones** are usually crystalline chemical precipitates. The principal minerals are calcite and dolomite. Soils formed from these rocks tend to be fine textured and near neutral or slightly alkaline in pH.

C. Metamorphic Rocks

Metamorphic rocks are formed by geological alterations of igneous, sedimentary or other metamorphic rocks which have been exposed to high heat and pressure deep in the earth for very long periods of time. They are important in certain locations, but are less common than sedimentary or igneous rocks.

1. **Slate** is a stratified metamorphic rock derived from shale. The mineral grains are extremely fine quartz and mica. Slate cleaves into sheets, but is more dense and compact than shale.

2. **Schist** has coarser crystals than slate, and consists mainly of wavy laminations (layers) of mica or other minerals often separated by irregular layers of granular quartz.

3. **Gneiss** (pronounced nice) is a coarse grained metamorphic rock derived from granite. It usually contains the same array of minerals. It is less stratified than schist.

4. **Marble** is metamorphosed limestone. Its major mineral is calcite or dolomite. Marbles weather easily, but more slowly than limestones.

5. **Quartzite** is an unstratified metamorphosed sandstone composed almost entirely of quartz. It is extremely resistant to weathering.

6. **Serpentinite** is a greenish colored, low-temperature metamorphic rock, rich in serpentine and other magnesium silicate minerals. It forms from the alteration of a group of igneous rocks known as ultramafic meaning high magnesium and iron. (Ultramafic rocks are not shown in the diagram for igneous rock classification, but would be to the right of gabbro.) Worldwide, serpentinite is not an important soil forming rock, but it is common in California. Serpentinite has been designated the state rock of California.

Procedure:

Complete this chart by placing an I for igneous, M for metamorphic or S for sedimentary rocks in the Class column. In the second column list the principal minerals contained in each rock. This information can be obtained from the descriptions above. In the last column briefly describe the main visual characteristics of each rock. Grain size is described as either macro-crystalline or microcrystalline. Coarse grained rocks contain minerals that are distinguishable with the unaided eye. In fine grained rocks, the individual minerals are not recognizable with the eye; nonetheless they may sparkle when viewed in bright light.

ROCK EXAMINATION CHART

ROCK	CLASS I,M,S	LIST PRINCIPAL MINERALS CONTAINED IN ROCK	DESCRIBE VISUAL CHARACTERISTICS A. COLOR B. GRAIN SIZE	
CONGLOMERATE				
BASALT				
GRANITE				
MARBLE				
LIMESTONE				
SANDSTONE				
SCHIST				
SHALE				
SERPENTINITE				

Rock Questions:

1. What is the dominant mineral in limestone?   _____

2. How can the difference between an igneous intrusive and igneous extrusive rock be recognized?

III. WEATHERING

(See Section 2.2 in the Text)

Weathering is the sum of the physical disintegration and chemical decomposition processes that changes rocks and minerals exposed to air and water.

Physical disintegration causes rock masses to split apart or to abrade. Freezing and thawing, abrasion by particles carried by wind or water and glaciers are factors of disintegration. The disintegration type of weathering precedes the decomposition process except for the more soluble rocks, such as limestones.

Chemical decomposition progresses most rapidly in warm moist environments. The smaller the particles, the greater the surface area exposed and the greater the rate of chemical decomposition. Several concurrent processes are involved in chemical decomposition: hydrolysis, hydration, carbonation, solution, and oxidation-reduction. This exercise will illustrates hydrolysis, carbonation, and oxidation-reduction.

Hydrolysis is a chemical reaction involving double displacement in which hydrogen of water combines with the anion of the mineral and hydroxyl of water combines with the cation of the mineral to form an acid and base. The mineral orthoclase illustrates the decomposition process.

$$KAlSi_3O_8 \quad + \quad HOH \longrightarrow HAlSi_3O_8 \quad + \quad K^+ + OH^-$$

Orthoclase   Water --->   Aluminum   +   Potassium and
                                           Silicate   Hydroxide ions

The potassium ion (K) released by this reaction is soluble and can be adsorbed by soil colloids, used by plants, or removed in the drainage water. What will the hydroxide ions (OH^-) do to the pH of the soil? The aluminum, silica, oxygen, and other ions may regroup to form a clay mineral such as kaolinite or smectite. This natural weathering process occurs at a very slow but continual rate in all soil parent materials and on rocks exposed to air and water.

Hydrolysis Procedure:

1. Rinse a 125 ml flask well with deionized water.

2. Place 8 drops of POLY-D pH Indicator and about 30 ml of deionized water into the clean flask.

3. Add about 0.5 grams of granitic sand into the flask, and swirl for about 30 seconds.

   What Is the pH of the mixture? _____

4. Empty the contents of the Erlenmeyer flask into the sink then rinse the flask with deionized water.

5. Repeat steps 2 and 3, but this time grind the sand to a very fine powder in a mortar.

    What is the pH of the pulverized sand and water mixture ? _____

6. Save this material for use in the carbonation exercise.

Hydrolysis Questions:

1. Does the change in pH indicate that the solution became more acidic or basic after grinding the sand? _____ Why did the pH change?

2. What is the effect of particle size on the rate of hydrolysis of minerals?

3. Why does particle size affect the rate of hydrolysis?

4. With all other factors being equal, how would the rate of soil formation in sand compare to the rate of soil formation in a silty parent material?

Carbonation and Solution

The metabolic activities of plant roots and organisms produce high CO_2 concentrations within soil pores. This CO_2 reacts with soil water to produce most of the carbonic acid found in soils. When carbon dioxide reacts with water, carbonic acid is produced:

$$CO_2 \quad + \quad H_2O, \text{ water} \longrightarrow H_2CO_3, \text{ carbonic acid}$$

The carbonic acid in water continually reacts with minerals as the water percolates through soils and rocks. Although, carbonic acid is a very weak acid, it is the most common acid found in soils.

$$H_2CO_3 \quad \dashrightarrow H^+ \quad + \quad HCO_3^-$$

$$\longleftarrow \text{-----------------}$$

Note: The short and long arrows show that the reaction tends to go more to the left than to the right, indicating that carbonic acid is a weak acid. The arrows would be of equal length for the dissociation of a strong acid.

Even though carbonic acid is a weak acid it is effective in accelerating the mineral decomposition processes as in the following reaction:

$$CaCO_3 + \qquad H_2CO_3 \quad \dashrightarrow \qquad Ca^{++} + 2HCO_3^-$$

Calcite   +   Carbonic acid ---> Soluble Calcium Bicarbonate

The calcium bicarbonate dissolves readily, releasing calcium ions (Ca^{++}) for plant uptake or possible leaching under excessive rainfall or irrigation. The loss of calcite from limestone leads to the development of acidic soils. The re-precipitation of calcite lower in the soil furthermore leads to the development of calcite hardpans. The formation of huge underground caverns results from the same process acting upon limestone rock below the soil.

Procedure:

1. Use the mixture of water, powdered minerals and POLY- D pH Reagent from the hydrolysis exercise. (Note: the granitic sand contains calcite.)

2. Use a sterile pipette to blow your breath into the mixture several times for about a minute. Record any changes in pH.

Initial pH: _____   Final pH _____

The following reactions take place in the flask:

$$H_2O \quad + \quad CO_2 \quad --------> \quad H_2CO_3$$

| Water | Carbon Dioxide | Carbonic Acid |

$$Then \quad H_2CO_3 + \quad CaCO_3 \quad ------> \quad Ca(HCO_3)$$

| Carbonic Acid | Calcium Carbonate ----> | Calcium Bicarbonate |

This reaction is significant in soils because the $Ca(HCO_3)_2$ produced is many times more soluble than $CaCO_3$ in the soil solution. Thus, the calcite becomes chemically weathered, releasing calcium ions for plant uptake. At the same time, the pH of the soil solution increases.

Carbonation and Solution Questions:

1. In what way did blowing your breath into the water affect the pH?

    _____

2. What do you predict will happen to the $CaCO_3$ (calcite) in soils that receive abundant rainfall or irrigation water annually?

    _____

3. List the sedimentary and metamorphic rock types most likely to develop into caverns by chemical weathering?

    _____   _____

Oxidation-reduction is a chemical weathering process involving the transfer of electrons from one reactant to another. The substance that looses electrons is oxidized and the substance that accepts the electrons is reduced. Commonly, but not always, free oxygen serves as the oxidizing agent(acceptor of electrons). It combines with an element in a mineral, such as iron (Fe), causing the iron atoms to give up some of their electrons to oxygen. The iron is oxidized and the oxygen is reduced. These electron transfers constitute a form of chemical weathering. Oxidation-reduction disrupts the atomic structure of minerals resulting the release of elements. The free elements often recombine to form new compounds including clay minerals.

$$2Fe^{++} + 2e^- + 3/2\ O_2 \ \text{------>} \qquad 2Fe^{+++} + 3O^{--} \qquad \text{------->} \qquad Fe_2O_3$$

| Reduced iron | Oxidized iron | Iron oxide |
| Black | Reddish | Reddish clay |

Although oxidation-reduction reactions take place continuously in soils, the rate can be very slow and unnoticeable. If a soil is moist and well aerated, iron tends to become oxidized (loses electrons to oxygen), but if air is absent, as in water-saturated soil, the iron tends to be reduced (gains electrons). The extent of either oxidation or reduction in soils is manifested in the color of the soil. Oxidized iron tends to produce red, yellow or brown colors in soils. Reduced iron, on the other hand, tends to produce black, gray, olive and blue colors. Alternating periods of oxidation and reduction caused by a fluctuating water levels in soil tend to cause **mottling** (blotches of red and gray colors). In some soils, ideal reducing conditions caused by prolonged water logging, tend to form gleyed conditions. These are noted by dark gray to black soil colors. A large population of anaerobic microorganisms in water logged, oxygen free soil speeds the reduction process.

Observe the colors of the gleyed soil on display and the soil monoliths assigned by your Instructor.

Oxidation and Reduction Questions:

1. What is the difference in appearance between the gleyed soil and mottled soil?

2. What causes the difference between a gleyed soil and a mottled soil?

3. Which soil horizon is most susceptible to reducing conditions? Explain.

4. In the field, describe a location where one would most likely find soils developing under reducing conditions.

5. Which soil horizon is most susceptible to oxidizing conditions? Explain.

6. What is the relationship between soil drainage and the oxidation-reduction process in soil?

7. Rocks containing iron-rich minerals tend to weather easily and become reddish brown when exposed to the atmosphere. Why?

8. Which soil condition, oxidized or reduced, would be best when selecting a site for the following soil uses?   Briefly explain.

   A   a place to install a home septic tank leach field to dispose of waste water.

   B.   a location to construct a house with a basement.

   C.   a place to grow perennial crops whose roots have high oxygen requirements.

   D.   an area to grow rice.

   E.   a site for a shopping center parking lot.

9. Transformation of rocks into parent materials and soil

   Fill in the following table after observing the rocks and soils in the trays on display.

   Note: the first tray contains the fresh rocks before being weathered, the second tray contains the parent material of the soil that formed from the rock, and the third tray contains the soil that has formed from the parent material.

Rock Type	Predominant minerals in rock	Changes in appearance between fresh rock and weathered rock material	Weathered rocks: Are iron rust stains apparent on the rocks?

LABORATORY EXERCISE II

Instructor copy

Name_____

Section _____

MINERAL EXAMINATION CHART

Mineral	Hardness hard, med, soft	Color	Cleavage yes or no	Plant nutrients released by weathering
Quartz				
Orthoclase				
Plagioclase				
Muscovite				
Biotite				
Hornblende				
Hematite				
Pyrite				
Apatite				
Calcite				
Gypsum				

Mineral Questions:

1. Which mineral is most resistant to weathering? _____

2. Name two feldspar minerals. _____ _____

3. Identify two micas. _____ _____

4. List two ferromagnesian minerals _____ _____

LABORATORY EXERCISE II

Instructor Copy

Name _____

Section _____

ROCK EXAMINATION CHART

ROCK	CLASS I,M,S	LIST PRINCIPAL MINERALS CONTAINED IN ROCK	DESCRIBE VISUAL CHARACTERISTICS A.COLOR	B. GRAIN SIZE
CONGLOMERATE				
BASALT				
GRANITE				
MARBLE				
LIMESTONE				
SANDSTONE				
SCHIST				
SHALE				
SERPENTINITE				

Rock Questions:

1. What Is the dominant mineral in limestone? _____

2. How can the difference between an igneous intrusive and igneous extrusive rock be recognized?

   _____

49

LABORATORY EXERCISE II

Instructor copy

Name _____

Section _____

Hydrolysis Questions:

1. Does the change in pH indicate that the solution became more acidic or more basic after grinding? Why did the pH change?

    _____

2. What effect does particle size have on the rate of mineral hydrolysis in water.

3. Why does particle size affect the rate of hydrolysis?

4. With all other factors being equal, how would the rate of soil formation in sand compare to the rate of soil formation in a silty parent material?

Carbonation and Solution Questions:

1. In what way did blowing your breath into the water affect the pH?

2. What happens to the $CaCO_3$ (calcite) in soils under high rainfall conditions?

3. Which rocks are most susceptible to being broken down by carbonation reactions?

LABORATORY EXERCISE II

Instructor copy

Name _____

Section _____

Oxidation and Reduction Questions:

1. What is the difference in appearance between the gleyed soil and mottled soil?

2. What causes the difference between a gleyed soil and a mottled soil?

3. Which soil horizon is most susceptible to reducing conditions? Explain.

4. Which soil horizon is most susceptible to oxidizing conditions? Explain.

5. What is the relationship between soil drainage and the oxidation-reduction process in soil?

6. Rocks containing iron-rich minerals tend to weather easily and become reddish brown when exposed to the atmosphere. Why?

7. Which soil condition, oxidized or reduced, would be best when selecting a site for the following soil uses? Briefly explain.

    a   a place to install a home septic tank leach field to dispose of waste water.

    b.   a location to construct a house with a basement.

    c.   a place to grow perennial crops whose roots have high oxygen requirements.

    d   an area to grow rice.

    e.   a site for a shopping center parking lot.

Instructor copy

Name _____

Section _____

8.   Transformation of rocks into parent materials and soil

Rock Type	Predominant minerals in rock	Changes in appearance between fresh rock and weathered rock material	Weathered rocks: Are iron rust stains apparent on the rocks?

Instructor copy

Name _____

Section _____

Assume that a perfect cube of basalt is broken into smaller cubes by successively splitting every cube into 10 equal parts along each side. Begin with a cube that is 10 millimeters (1 cm) on an edge. Assume perfect cleavage with no waste in splitting. Calculate the total number of particles, the total length of exposed edges and the total amount of exposed surface area on all the particles in each group. <u>Use dimensional analysis</u>. Fill in the chart with your answers and neatly show all calculations on separate pages of notebook paper. Attach your work to the back of this page.

name of particle group	length of one exposed edge	total number of particles	total length of exposed edges	total exposed surface area
	mm		meters	cm^2
gravel	1×10^1	1	1.2×10^{-1}	
coarse sand	1×10^0	1×10^3		
fine sand	1×10^{-1}			600
silt	1×10^{-2}			
clay	1×10^{-3}	1×10^{12}		
colloidal clay	1×10^{-4}			600,000

What is the length of all of the edges on the colloidal particles in miles and kilometers ?

miles _____   kilometers _____

The 1.0 cubic cm block of basalt weighing 2.7 grams is pulverized to silt size and transferred to a rectangular container 10 mm on each side of the base. The silt-sized particles filled the container to the 25 mm mark because the particles can no longer fit perfectly together. Calculate the density of the whole bulk pile of silt-sized particles in grams per cubic centimeter.

_____ (remember to include units)

LABORATORY EXERCISE III

SOIL TEXTURE, STRUCTURE AND WATER RELATIONS

Goal: To understand soil texture and structure and their influence on soil water relations.

Objectives:

1. Understand the definition of all words used in this exercise.

2. Define soil texture and understand its importance.

3. Learn to calculate % sand, % silt, and % clay, given the appropriate data.

4. Learn to use the soil textural triangle.

5. Determine soil texture by feel.

6. Define soil structure and understand its importance.

7. Recognize various types of soil structure.

8. Understand the relationships between soil texture and structure.

9. Understand the influence of soil texture and structure on water infiltration and percolation.

10. Understand the influence of soil texture and structure on soil water holding capacity.

11. Determine moisture content of a soil sample by the gravimetrics method, and learn to use the % moisture equation.

12. Explain why the oven dry weight is **always** used when calculating percent moisture in mineral soils.

LABORATORY EXERCISE III

Soil Texture, Structure and Water Relations
(See Section 3.2 pages 70-74 in the Text)

This is an integrated exercise: it combines soil texture, structure and water relations. You will complete the texture and structure parts today, but will continue measurements on the water portion for two more weeks.

First, you will observe examples of soil textures and structures independently; second, you will observe the effects of texture and structure on water percolation; third, you will measure the water holding capacity of a coarse and fine textured soil.

PART 1. SOIL TEXTURE AND STRUCTURE

The solid portion of soils consists primarily of mineral particles (usually more than 95% by weight) mixed with organic materials. The mineral particles are divided into **coarse fragments**, greater than 2mm in diameter (e.g., gravel), and **fines**, less than 2mm in diameter. The fines are further divided into arbitrary size classifications known as soil **separates**. The three soil separates are sand, silt, and clay. The size classification of these soil separates is shown in Table 7.

Table 7. Size Classification of Soil Separates*

Separate	Diameter (mm)
sand	2.0 to 0.05
silt	0.05 to 0.002
clay	less than 0.002

*U.S. Dept. of Agriculture System of Soil Separate Classification
See Table 3-1 in the text, pp 71.

Soil **texture** is defined by the relative proportion of sand, silt, and clay. (It applies only to mineral soils; organic soils such as peat and muck, do not have texture.) Because texture is limited to sand, silt and clay, other soil components, including water, air, organic matter, and inorganic materials larger than 2mm, are excluded. However, gravels and cobbles may be included as modifiers to the textural name. The percentages of sand, silt and clay are calculated from laboratory data according to the following example:

(Remember that % is defined as parts per hundred.)

Example:

mass of dry, inorganic soil less than 2 mm diameter = 50.0 g

mass of sand   = 15.5 g
mass of silt   = 17.0 g
mass of clay   = 17.5 g

% sand = $\frac{15.5 \text{ g}}{50.0 \text{ g}}$ = $\frac{0.31 \text{ g}}{1.0 \text{ g}}$ = 31.0 g/100g = 31% or 31 pph

% silt = $\frac{17.0 \text{ g}}{50.0 \text{ 9}}$ = $\frac{0.34 \text{ g}}{1.0 \text{ g}}$ = 34.0 g/100g = 34% or 34 pph

% clay = $\frac{17.5 \text{ g}}{50.0 \text{ 9}}$ = $\frac{0.35 \text{ g}}{1.0 \text{ g}}$ = 35.0 g/100.0g = 35% or 35 pph

Soils having similar textures can be grouped into classes of soils having similar characteristics, behavior, and management requirements. There are 12 soil textural classes as shown in the textural triangle on the next page. The textural classes provide a way to express soil texture without referring specifically to the percentages of each separate. When any two of the three percentages are known, the soil textural class can be determined from the triangle. The percent sand, silt, and clay must always total 100% for a sample.

What is the textural class for the soil in the example above?

clay loam or silt loam

GUIDE FOR SOIL TEXTURE CLASSIFICATION

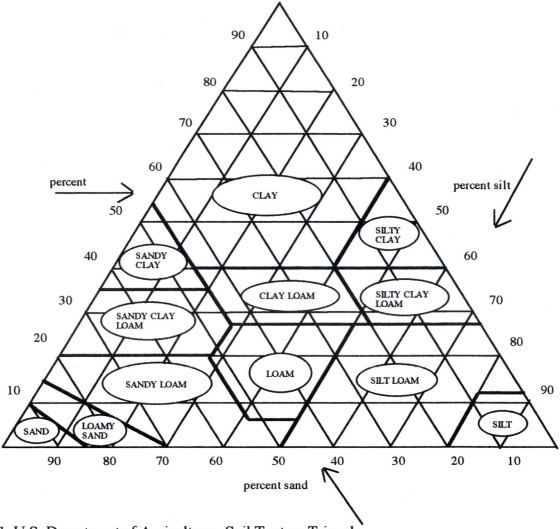

Figure 1. U.S. Department of Agriculture: Soil Texture Triangle
(Drawing by Matthew Ballmer, 1998)

A soil's textural class can be determined by several methods. They range from quick and simple field tests to more involved laboratory techniques. Laboratory methods are used to determine the exact particle size distribution, and are known as **particle size analyses**, or, sometimes, **mechanical analyses**. The methods vary in the amount of time, equipment required, and in accuracy, but they all are based on principles of particle separation by sedimentation.

The "field" or "feel" method is the simplest to conduct but its accuracy depends on the talent and experience of the investigator. The "feel" method, is qualitative in nature. It involves moistening the soil, kneading it with the fingers to mix thoroughly, and feeling the particles to determine the textural class. The clay content in the sample is estimated by how well the soil ribbons. The accuracy of estimating soil texture by this method obviously improves with experience.

Texture is a very important soil property, because it influences most other soil properties. As the clay content of a soil increases so does the surface area. One hand full of a clay textured soil will have a total exposed surface area of about five acres. A hand full of sand however, may have only about one acre of exposed surface area to hold water and nutrients. Note: **coarse textured soils** are high in sand, **fine textured soils** are high in clay.

Table 8. A general comparison of some soil properties for coarse and fine textured soils.

	Coarse textures	Fine textures
Structure	generally structureless	granular, blocky or prismatic
Porosity	lower	higher
Bulk density	higher	lower
Tilth	easy to till	difficult to till
Water infiltration	fast	slow
Water holding	low	high
Nutrient holding	low	high
Shrink/swell	low	high

Laboratory Exercise IV
Routine Soil Particle Size Analysis
(Optional)

The proceedure for separating soil into various size groups is called **particle-size analysis**, or **mechanical analysis**. Usually the soil is first separated into material larger than two millimeters from particles smaller than two millimeters to remove roots, gravel, and rubble. The material that is less than two millimeters is next treated to remove humus by chemical oxidation using hydrogen peroxide. **Routine mechanical analysis**, to obtain (approximate) textural classes, are done without the removal of humus. Knowing the texture of a soil helps one make wise decisions about the use and management requirements for a soil. Cultivating and irrigating a clay or a sand is vastly different.

The basis for determining the texture of a soil is **Stokes Law** of particle settling velocities in water. The simplified version of Stokes Law is written as follows:

$$V = 8711D^2 \text{ cm/sec}$$

Where V is in cm/sec and D is the diameter of particles in cm.

The number 8711 is a constant obtained by combining the other factors in Stokes Law. Factors including particle density, temperture, water density, gravity and viscosity are not variable under a standard condition of 20°C at the earth's surface.

A **hydrometer** inserted into a water suspension of soil particles measures the grams of particles in the top 10 cm of the mixture. Solving Stokes Law for the settling velocities of sand and silt produces the time required for determining when to insert the hydrometer to obtain the correct readings used for soil texture analysis.

The time for medium sand(0.25mm) to settle 10 cm can be found as follows:

$$V = 8711(.025)^2 \quad \text{(remember particle size must be in cm)}$$

$$V = 5.4 \text{ cm/sec}$$

In 1.8 seconds the 0.25 mm sand particles will have settled 10 cm in the suspension

Proceedure for a Routine particle-size analysis:

This exercise may be done individually, as a group, or as a demonstration by the instructor.

1. Obtain 50.0 grams of dry soil that is less than two millimeters in size

2. Transfer the soil to a soil blender cup and add 25 ml of a 10% sodium hexametaphosphate solution. Sodium hexametaphosphate is a dispersing agent that causes the sand, silt and clay particles to separate.

3. Add 250ml of deionized water (DI)to the cup and blend for two minutes.

4. Stop the blender, remove the cup, and transfer all the suspension into a Bouyoucos cylinder. Use a squeeze bottle and DI water to remove all of the particles from the cup.

5. Fill the Bouycous cylinder to the 1000ml mark with DI water.

6. Stir the suspension and gently insert the hydrometer into the mixture after 30 seconds.

7. Steady the hydrometer from bobbing up and down and read, on the hydrometer stem, the grams of silt and clay particles in the suspension at exactly 45 seconds.

   Note: very fine sand particle size ranges from 0.1 - 0.05 mm in diameter

   $$V = 8711D^2 \qquad 8711(0.005)^2 = 0.22 \text{ cm/sec}$$

   It will take the 45 seconds for all of the fine sand to settle out of the top 10 cm of the suspension.

8. Remove the hydrometer and return it to its case to prevent breakage. Set the cylinder aside to allow the silt particles to settle before the next reading.

9. Determine, using Stokes law above, the time required for the 0.005 mm soil particles to settle out of the top 10 cm of the solution. Check your answer with the instructor before completing the next hydrometer reading.

   Time required for the 0.005 mm silt particles to settle 10 cm. _____
   Show calculations here and verify results with the instructor

10. Complete the section on Texture by Feel on pp 69-70.

11. Routine Particle-sise Analysis Results:

    Soil Sample _____

    A. Grams of dry soil in the Bouyoucos cylinder. _____ g

    B. Hydrometer reading after 45 seconds. _____ g

    C. Grams of silt and clay in suspension. _____ g

    D. Grams of sand in the sample. _____ g

    E. Grams of clay in the sample. _____ g

    F. Percent sand, silt, and clay in the sample. _____ . _____ . _____

    F. Soil Texture Class _____

Questions on Soil Texture:

1. How would the water temperature affect the falling velocity of small soil particles in a suspension? (Hint: which factors in Stokes Law are affected by temperature)

2. Find the time required for large clay (0.002 mm) particles to settle to the bottom of a three meter deep pond using Stokes Law.

3. What effect could the humus left in the 50 gram soil sample above have on determining the soil's texture?

4. The clay content of soils affects several soil properties. On the generalized graphs below, place the letter corrosponding to each soil property on the left side of graph A or B that best illustrates the relationship of the soil property to increasing clay content in soil.

   a. swelling potential   b. soil drainage ease   c. nutrient holding ability
   d. structure formation   e. ease of compaction   f. % humus
   g. ease of tilling   h. landscaping top soil   i. dry mass per cubic foot

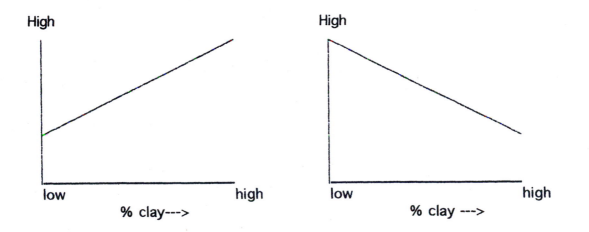

5. Complete the section on Texture and Structure, pp 70-72 below.

SELECTED SOIL TEXTURE CLASS CHARACTERISTICS

(The following are suggestive only and all may not occur with any particular soil because of differences in clay types, organic matter content, exchangeable cation ratios, or amount of soluble salts present.)

SAND OR LOAMY SAND: Dry-loose, single grained; gritty, no or very weak aggregates. Moist - gritty; forms easily crumbled ball; does not ribbon. Wet - lacks stickiness; may show faint clay staining (loamy sand, especially). Individual grains can be both seen and felt under all moisture conditions.
SANDY LOAM: Dry - aggregrates break easily. Moist - moderately gritty to gritty; forms ball that stands careful handling; ribbons very poorly. Wet - definitely stains fingers; may have faint smootheness or stickiness, but grittiness dominates. Individual grains can be seen and felt under nearly all conditions.

LOAM: This is the most difficult texture to place since characteristics of sand, silt, and clay are all present but none predominates. Suggests other textures. Dry - aggregates slightly difficult to break; somewhat gritty. Moist - forms firm ball; ribbons poorly; may show poor fingerprint. Wet - gritty, smooth, and sticky all at the same time.

SILT OR SILT LOAM: Dry - aggregates moderately difficult to break and ruptures suddenly to a floury powder that clings to fingers; shows fingerprint. Moist - has smooth, slick, velvety, or buttery feel; forms firm ball; may ribbon slightly before breaking; shows good fingerprinting. Wet - smooth with some stickiness from clay. Grittiness of sand is well masked by other separates.

CLAY LOAM: Dry - clods break with difficulty. Moist - forms a firm ball that dries moderately hard; ribbons fairly well, 1-2 inches, but ribbons barely supports own weight; shows fair to good fingerprint. Wet - moderately sticky with stickiness dominating over grittiness and smoothness; stains fingers.

CLAY: Dry - cloddy, aggregates often cannot be broken even with extreme pressure. Moist - forms firm, easily molded ball drying very hard; squeezes out to a very thin ribbon, 2-3 inches long. Wet - stains fingers, usually very sticky with stickiness masking both smoothness and grittiness; wets slowly.

SANDY CLAY: - Dry - often cloddy, aggregates broken with extreme pressure. Moist - forms very firm ball, drying quite hard; squeezes to a thin long gritty ribbon. Wet - stains fingers; clouds water, sticky and plastic with grittiness.

SILTY CLAY LOAM: Resembles SILT LOAM but with more stickiness of clay. Dry - aggregates break with great difficulty. Moist - shows a good fingerprint; forms a firm ball drying moderately hard; ribbons 1/2 inch; Wet - stains fingers; has sticky smooth feel with little grittiness of sand.

SILTY CLAY: Dry - cloddy, aggregates broken with extreme pressure. Moist - forms a very firm ball becoming very hard upon drying; shows fingerprint; squeezes out to a long smooth ribbon. Wet - stains finger, clouds water; stickiness dominates over smoothness, grittiness absent.

Determining Soil Texture by "Feel"

The soil textural class can be estimated by observing and feeling the soil under dry, moist and wet conditions. The size range of the separates and their feel when MOIST are listed below:

Table 9. Soil separate, size range and feel when moist.

Separate	Size Range	Feel
Sand	2.0 to 0.05 mm	gritty
Silt	0.05 to 0.002 mm	smooth, floury
Clay	less than 0.002 mm	buttery, sticky, plastic, hard when dry

PROCEDURE:

1. Obtain four soils of known textures from the window bench. Place about two tablespoon of each on a paper towel and carry them back to your seat. Be sure to write the soil textural class on the paper towel.

2. Examine the DRY soil for aggregates and the ease of breaking them by pressing between your fingers. Soils high in sand are seldom strongly aggregated. Soils high in silt may be aggregated, but usually break suddenly into a soft powder. Soils high in clay are usually strongly aggregated. Dry aggregates are often hard to break, even with extreme pressure.

3. Moisten the soil gradually with <u>small</u> <u>amounts</u> <u>of</u> <u>water</u>, kneading it vigorously until all dry lumps are wetted. (If too much water is used, increase the volume of soil.) When all of the soil is moist, test the ease of forming a ball, then try to squeeze the soil into a ribbon by working it between your thumb and forefinger. See the pictures at the front lab bench.

4. a. Sand, sandy loams, and loamy sands will not ribbon
   b. Loams, and silt loams form ribbons less than one inch long
   c. Sandy clay loam, clay loam, and silty clay loams will form 1 - 2 inch ribbons
   d. Sandy clay, clay, and silty clay will from ribbons greater than two inches

   1. Soils high in sand feel gritty, and ribbon poorly except when also high in clay. You usually will be able to see the sand grains - look for them. Sand gives a grinding sound when rubbed between the fingers. Generally, when sand can be felt in a moist sample there is at least 35% sand present.

   2. Soils high in silt feel smooth, velvety, or floury. They may form a short ribbon, the length of which varies with the clay content. Dark soils high in organic matter feel more silty than analysis shows them to be, because humus usually feels like silt.

   3. Soils high in clay can often be pressed out into very thin ribbons two or more inches long.

5. Record your observations on the table below.

6. Repeat the procedure steps 1 through 5 for the unknown samples, and estimate the textural class of each.

A.   **Known Samples**

   Soil Characteristics   Soil Textural Class

   (grittiness, stickiness, ribbon length)

   1. _____   _____

   2. _____   _____

   3. _____   _____

   4. _____   _____

B.   **Unknown Samples**

   Soil Characteristics   Soil Textural Class

   (grittiness, stickiness, ribbon length)

   1. _____   _____

   2. _____   _____

   3. Desk Soil_____   _____Number ___

TEXTURE QUESTIONS:

1.   Define soil texture.

2.   Place on the graph a line that shows the general relationship between surface area and soil texture for an equal mass of sand, silt loam, and clay samples.

```
High  |
      |
Surface
Area  |
      |
      |
      |
Low   |__|_____|_____|_____
         sands  silt loam  clays
```

3. What is the lowest clay percentage allowed for a texture to be classed as:

   a. clay _____   c. clay loam _____

   b. sandy clay loam _____   d. sandy clay _____

4. What is the lowest silt percentage allowed for a texture to be classed as:

   a. loam _____ b. silt loam _____ c. sandy loam _____

5. Practice determining textural class of soils.

% sand	40		10	22	40
% silt	40	25		60	46
% clay	20	25	20	18	

   Textural class _____   _____   _____   _____   _____

6. A dry soil sample weighing _____grams contains _____grams sand,_____ grams clay, and the remainder is silt. Calculate the percentages of sand, silt and clay in the soil and determine its textural class. Show the appropriate units and complete setups for the problem.

   % sand _____

   % silt _____

   % clay _____

   Texture class _____

7. A moist soil sample containing _____grams of water, and _____grams of organic matter weighs _____grams. It contains _____% sand, _____grams silt, and the remaining mineral fraction is clay. Calculate the grams of sand, the percent silt, and the grams and percent clay in the sample. What is the soil's textural class? Show the appropriate units and complete setups for the problem.

   grams sand _____

   % silt _____

   % clay _____

   Texture class _____

8. Which soil separate will settle most rapidly to the bottom of the lake?

   _____

9. Which soil separate contributes most to water *turbidity? *(muddy water)

   _____

10. Which soil separate is the subject of a whole field of study in Soil Science?

    separate _____Field of study _____

SOIL STRUCTURE
(See Section 3.4 pages 75-78 in the Text)

Soil structure is the arrangement of the individual soil particles into aggregates separated by surfaces of weakness. These structural aggregates, called **peds**, comprise several types depending on their shape: **crumb, granular, blocky, prismatic, columnar, and platy.** Note that the definition has two parts: aggregation and separations. If a soil's condition does not satisfy both parts of the definition, it is structureless. The two structureless conditions are single grained (no aggregation) and massive (no separation patterns).

Water infiltration, percolation, aeration, erosion resistance, and ease of root penetration are influnced by a soil's structure. Texture, soil chemistry , humus content, plants, animal and human activity all interact to create different soil structure conditions. The common cementing agents holding the separates together to form peds include clay, humus, silica, lime, and iron oxides.

Soil structure and texture interact to control other soil properties.

Note in Table 7 that soil structure has greater influence in fine textured soils than in coarse textured soils. This indicates that structure is more important in fine soils than in coarse soils.

PROCEDURE:

1a. Draw a sketch of the five soil structures on display.

Structure Type:_____ _____ _____ _____ _____

1b. The structureless conditions are _____ _____

2.   Observe the monoliths on display to complete the following:.

   Which monolith has:

   a.   a soil with granular structure?   _____

        In which horizon does that structure occur? _____

   b.   a soil with prismatic structure?   _____

        Which horizon has prismatic structure?   _____

   c.   a soil with blocky structure?   _____

   d.   a soil that is mostly structureless?   _____

PART 11. WATER FLOW THROUGH SOILS, and WATER HOLDING CAPACITY
(See section 4.7 pages 129-133 in the Text)

Several factors control the movement of water through soils. These include soil texture, structure, organic matter content, impervious layers, water content, temperature, and compaction. The use of soils for agriculture, sewage disposal, and waste management makes it necessary to understand the physical properties of soils that control both infiltration and percolation.

In this exercise, you will be recording the amount of water percolating through one coarse and two fine textured soils having different structural conditions. The soils' textural class are a sandy loam and two silt loams. The sand is structureless. One sample will have granular structure and the other will have the soil's structure destroyed by crushing.

Equipment Needed:
  funnel rack and stand -1
  percolation tubes - 3
  100 ml graduated cylinder -4
  cheesecloth - 3 pieces
  250 ml beaker 3
  timer
  moisture cans - 3

Procedures:

Note: Students are to work in groups as assigned by the instructor.

1.   Obtain three percolation tubes. Label them A, B, and C. Place one piece of cheesecloth at the bottom of each percolation tube to prevent soil from clogging the stem. (You may need to add a few drops of water to hold the cheesecloth in place.)

2.   Fill each tube 2/3 full with the following soils: Tube A - sandy loam, Tube B -well granulated clay loam, Tube C - the clay loam after destroying structure by grinding in a mortar.

3.   Place each tube with soil in the funnel rack, use a 100 ml graduated cylinder below the stem of each percolation tube to catch the water.

4.   Add water with a beaker to tube A to saturate the soil. Maintain the water level one cm above the soil surface for the duration of the experiment.

      Begin timing when the first drop of water flows from the tube.

      Record on the data sheet the amount of water that comes through each minute for the first 10 minutes. Continue the measurements to obtain a 20 minute reading. If needed use a second cylinder to catch volumes of water greater than 100 ml. After the last reading, replace the 100 ml graduated cylinder with a 250 ml beaker, and allow the excess water to drain. This excess water will be discarded. **Save the soil for step 6.**

5.   Repeat step 4 for the soils in percolation tubes B and C.

6.   Obtain three numbered soil moisture cans. Brush out and weigh to the nearest 0.1 gram each can and lid. Record the number and weight of each can and lid on the soil moisture data page in Part III PP 79.

7. Transfer the saturated soil from the percolation tubes to the soil moisture cans. Remove and discard the piece of cheesecloth. Pour off excess water. Be sure to record which soil is in each can.

8. Weigh and record the mass of soil can and lid to the nearest 0.1 g. Record the data on the soil moisture data page 79. The soil mass is assumed to be near saturation.

9. Place the lid on the bottom of the can. Put the three cans of soil on the storage tray as indicated by the Instructor. The soil will be allowed to air dry in a warm room for one week.

10. Clean your work area and equipment then plot the percolation data on the graph provided with the data sheet. Answer the questions.

11. After one week of drying the soil and cans are to be weighed. Record the new weight as the air dry weight on the soil moisture data page. The cans are ready to be placed in a drying oven at 105 Celsius for 24 hours.

12. After another week, weigh the can and oven dry soil. Record the results on the soil moisture data page, 79.

13. Empty the cans of dry soil into the trash, clean the can inside with a dry paper towel, replace the lid on the can, and return the cans to the storage drawers.

14. Complete the calculations on the soil moisture data page and answer the questions.

Name _____

Section _____

WATER PERCOLATION DATA PAGE

Soil Texture and Structure Condition

MINUTES	Sandy loam Cumulative ml	ml each minute	Granular Clay loam Cumulative ml	ml each minute	Pulverised Clay loam Cumulative ml	ml each minute
1						
2						
3						
4						
5						
6						
7						
8						
9						
10						
15						
20						

Note: The ml/min is determined by finding the percolation rate for each minute from the cumulative data.

Directions: Neatly plot all three cumulative curves on one sheet of paper using a computer. (A sheet of regular graph paper can be found on the last page in this manual). Answer the following questions from the data and attach the graph to the Instructor Copy, page 89.

WATER MOVEMENT QUESTIONS

1. Did the saturated soil percolation rate remain constant over time for each of the soils in this experiment? Explain why or why not.

2. What effect did soil texture have on the rate of water percolation for each treatment?

3. What effect did soil structure have on the rate of water percolation for each treatment?

4. On a 15% slope and under the same vegetative cover, which of the above soils would be most susceptible to erosion by water? Explain.

5. What soil property determines the sprinkler irrigation set time for the efficient application of water to crop land?

Part III. SOIL MOISTURE RELATIONSHIPS
(See Section 4:8 pages 134-136 in the Text)

Plants use several mechanisms to obtain water from soil. **Passive absorption** is the most important and accounts for more than 90 percent of the water absorbed by plants. (See the text for additional mechanisms plants use to obtain water) Passive absorption occurs as a result of the pulling force created in a plant's vascular system as water is lost at the leaf surface by transpiration. When a water molecule leaves the plant leaf surface other molecules move upward to take its place. As the water molecules in the plant move upward they pull those in contact with the plant root inside. The more water plants remove from a soil the more difficult it becomes to obtain the remaining water. Plants must over come the forces of **adhesion** and **cohesion** in order to obtain water from soil. Plants are able to use only a portion of the total water held in the soil pore space, as illustrated in Figure 2.

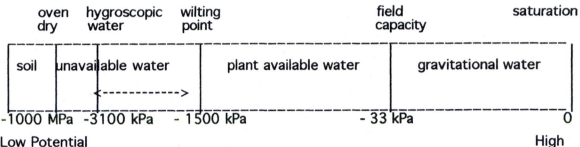

Figure 2. Water potential equivalents in kilopascals[*] for water held in the pore space of soil.

Equivalent force, in kiloPascals, plants

need to overcome to obtain water from soil

------> increasing amount of water in soil - - - - - - - - - - - - - - - - - - - >

Note: a Pascal is a unit of pressure. One Pascal equals the force of one Newton/square meter. 1 kilopascal = 1000 pascals, 1Mpa = 1 million pascals

Water moves through a soil in the large pores and drains away in response to force of gravity . This drainage water is GRAVITATIONAL WATER. Water held within small soil pores by adhesion and cohesion is CAPILLARY WATER. Only a portion of the water in the small soil capillaries is available to plants. AVAILABLE WATER is the water plants can obtain from the soil. It is the difference between the water content in soil at FIELD CAPACITY and the soil water content at PERMANENT WILTING POINT. For most common field crops, the permanent wilting point value is equated to -1500 kPa of water potential. GRAVITATIONAL WATER represents the difference in soil moisture content between SATURATION and FIELD CAPACITY. The water in air dry soil under conditions of 98 percent relative humidity is held as HYGROSCOPIC WATER(a film of water around soil particles). Soils that have been dried at 105 °C to a constant weight are said to be OVEN DRY. The oven dry weight of soil is the fixed reference weight used to quantify the amount of water in mineral soils.

(See section 4:3, pages 110-122 in the Text)

The basic equations used to quantify soil moisture are:

$$\text{Eq. 1} \quad \begin{array}{c} \% \text{ H2O} \\ \text{(by weight)} \end{array} = \frac{\text{(Wet weight soil - Oven dry weight soil)}}{\text{Oven dry soil weight (g)}} \times 100$$

$$\text{Eq. 2} \quad \begin{array}{c} \% \text{ H2O} \\ \text{(by volume)} \end{array} = \begin{array}{c} \% \text{ H2O} \\ \text{(by weight)} \end{array} \times \frac{\text{soil bulk density}}{\text{density water}}$$

$$\text{Eq. 3} \quad \begin{array}{c} \text{Depth of water} \\ \text{(in soil)} \end{array} = \frac{\text{soil bulk density}}{\text{density of water}} \times \frac{\text{percent water}}{100\%} \times \text{soil depth}$$

To illustrate the use of the above equations, let's assume a 100 gram sample of moist soil loses 25 grams by oven dry ing. From Eq.1 the moist sample contained 25/75 = 33/100 or 33% moisture. If the soil has a bulk density of 1.30g/cm^3 , the depth of water in 30 cm of the moist soil is 12.87 cm. Can you verify the results using Eq. 3?

SOIL MOISTURE DATA PAGE

Treatment	A	B	C
Soil Series name	_____	_____	_____
Soil textural class	_____	_____	_____
Soil structure	_____	_____	_____
Can number	_____	_____	_____
Weight of can and lid	_____	_____	_____
Weight of can, lid and saturated soil	_____	_____	_____
Weight of can, lid and air dry soil	_____	_____	_____
Weight of can, lid and oven dry soil	_____	_____	_____
Weight of oven dry soil	_____	_____	_____

Calculate the percent water in each sample for the following conditions, and show the math for your calculations in the space below:

1) when saturated _____ _____ _____

2) at air dry conditions _____ _____ _____

3) the cm of water in 30 cm _____ _____ _____
   of soil when saturated (assume
   bulk density value of 1.48, 1.30,
   and 1.39 g/cm^3 for A, B, C respectively)

SOIL MOISTURE QUESTIONS:

1. Which soil contained the most amount of water at saturation? Explain.

2. Which soil contained the most amount of water at air dry? Explain.

Soil moisture problems:

1. A soil contains _____ percent water when saturated, _____ percent water at field capacity, _____ percent water at the wilting point, and _____ percent water at air dry. The soil weighed _____ pounds per cubic foot when oven dry. (Instructor will provide values)

    a. What is the percent water available to plants at field capacity?

                                                    _____

    b. How many inches of water are available for plants per foot depth of soil at field capacity?

                                                    _____

    c. If the evapo-transpiration rate for a crop is 0.2 inches per day and irrigation is to be done when half the plant available water is used, find the time between irrigations.

                                                    _____

2. A wet sample of soil weighs 277.9 grams and contains 38.7 % moisture. Find the sample's oven dry weight.

                                                    _____

3. A planter bed 8 ft x 10 ft x 2 ft was filled with dry soil described at the top of this assignment. If the soil is to be brought to field capacity, find the gallons of water that needs to be added to the bed.

                                                    _____

4. What mechanisms do plants use to obtain water from soil?

LABORATORY EXERCISE III

Name _____

                                                               Section _____

Texture by Feel:

A   Known Samples

        Soil Characteristics

           (grittiness, stickiness, ribbon length)   Soil Texture

1.   _____   _____

2.   _____   _____

3.   _____   _____

4. _____   _____

B.   Unknown Samples

        Soil Characteristics

     (grittiness, stickiness, ribbon length) Soil Texture

1.   _____   _____

2.   _____   _____

3.   _____   _____

TEXTURE QUESTIONS:

1.   Define soil texture.

2.   Place on the graph a line that shows the general relationship between surface area and soil
     texture for an equal mass of sand, silt loam, and clay samples.

```
    High |
         |
 Surface |
   Area  |
         |
         |
         |
     Low |  ___|_____|_____|____
            sands     silt loam     clays
```

3.   What is the lowest clay percentage allowed for a texture to be classed as:

     a. clay   _____   c. clay loam _____

     b. sandy clay loam   _____   d. sandy clay _____

4. What is the lowest silt percentage allowed for a texture to be classed as:

   a. loam _____ b. silt loam _____ c. sandy loam _____

5. Practice determining textural class of soils.

% sand	40		10	22	40
% silt	40	25		60	46
% clay	20	25	20	18	

   Textural Class _____ _____ _____ _____ _____

6. A dry soil sample weighing _____grams contains _____grams sand,_____ grams clay, and the remainder is silt. Calculate the percentages of sand, silt and clay in the soil and determine the its textural class. Show the appropriate units and complete setups for the problem.

   % sand _____

   % silt _____

   % clay _____

   Textural class _____

7. A moist soil sample containing _____grams of water, and _____grams of organic matter weighs _____grams. It contains _____% sand, _____grams silt, and the remaining mineral fraction is clay. Calculate the grams of sand, the percent silt, and the grams and percent clay in the sample. What is the soil's textural class? Show the appropriate units and complete setups of the problem.

   grams sand _____

   % silt _____

   % clay _____

   Textural class _____

8. Which soil separate will settle most rapidly to the bottom of the lake?

   _____

9. Which soil separate contributes most to water turbidity ? (muddy water.)

   _____

10. Which soil separate is the subject of a whole field of study in Soil Science?

    separate _____Field of study _____

Name _____

Section _____

Directions: Neatly plot all three cumulative curves on one sheet of paper using a computer. (A sheet of regular graph paper can be found on the last page in this manual). Answer the following questions from the data and attach the graph to the Instructor Copy, page 89.

WATER MOVEMENT QUESTIONS

1. Did the percolation rate remain constant over time for each of the soils in this experiment? Explain why or why not.

2. What effect did soil texture have on the rate of water percolation for each treatment?

3. What effect did soil structure have on the rate of water percolation for each treatment?

4. On a 15% slope and under the same vegetative cover, which of the above soils would be most susceptible to surface erosion? Explain.

5. What soil parameter determines the sprinkler irrigation set time for delivering an efficient application of irrigation water?

SOIL STRUCTURE QUESTIONS

INSTRUCTOR COPY   Name _____

                                                                  Section _____

1. Define soil structure.

2. What major characteristic distinguishes granular structure from blocky structure?

3. How are soil texture and structure affected by soil compaction?

4. Name five cementing agents important in the formation of soil structure:

   _____   _____   _____   _____   _____

5. Which structure is best for a home garden? Why?

6. How would the water infiltration rate into a clayey soil be affected by soil structure?

7. Which soil structure is commonly found in A horizons of undisturbed grassland soils? Explain. _____

8. Identify a soil monolith and horizon with the following structure conditions:

   a. a monolith with granular structure   _____   horizon _____

   b. a monolith with prismatic structure _____   horizon _____

   c. a monolith with having no structure _____

9. Soil structure can be greatly influenced by soil management practices. List two that tend to improve soil structure and two that destroys structure.

   Structure promoting practices   **Structure destroying practices**

   1. _____   1. _____

   2. _____   2. _____

LABORATORY EXERCISE 3

Name _____

Section _____

WATER PERCOLATION DATA PAGE

Soil Texture and Structure Condition

MINUTES	Sandy loam		Granular Clay loam		Pulverised Clay loam	
	Cumulative ml	ml each minute	Cumulative ml	ml each minute	Cumulative ml	ml each minute
1						
2						
3						
4						
5						
6						
7						
8						
9						
10						
15						
20						

LABORATORY EXERCISE III

INSTRUCTOR COPY

Name _____

Section _____

SOIL MOISTURE DATA PAGE

Treatment	A	B	C
Soil texture	_____	_____	_____
Soil structure condition	_____	_____	_____
Can number	_____	_____	_____
Weight of can and lid	_____	_____	_____
Weight of can, lid and saturated soil	_____	_____	_____
Weight of can, lid and air dry soil	_____	_____	_____
Weight of can, lid and oven dry soil	_____	_____	_____
Weight of oven dry soil	_____	_____	_____

Calculate the percent water in each sample for the following conditions, and show the math for your calculations in the space below:

1) when saturated   _____   _____   _____

2) at air dry conditions   _____   _____   _____

3) the cm of water in 30 cm _____   _____   _____
    of soil when saturated (assume
    bulk density value of 1.48, 1.30,
    and 1.39 g/cm^3 for A, B, C respectively)

SOIL MOISTURE QUESTIONS:

1. Which soil contained the most amount of water at saturation? Explain.

2. Which soil contained the most amount of water at air dry? Explain.

Soil moisture homework: Name: _____

1.   A soil contains _____ percent water when saturated, _____ percent water
        at field capacity, _____ percent water at the wilting point, and _____
        percent water at air dry. The soil weighed _____ pounds per   cubic foot when
        oven dry. (Instructor will provide values)

    a.   What is the percent water available to plants at field capacity? (Show calculations)

                                                            _____

    b.   How many inches of water are available for plants per foot depth of soil at
            field capacity?

                                                            _____

    c.   If the evapo-transpiration rate for a crop is 0.2 inches per day and irrigation is
            to be done when half the plant available water is used, find the time between
            irrigations.

                                                            _____

2.   A wet sample of soil weighs 277.9 grams and contains 38.7 % moisture. Find the
        sample's oven dry weight.

                                                            _____

3.   A planter bed 8 ft x 10 ft x 2 ft was filled with dry soil described at the top of this
     assignment. If the soil is to be brought to field capacity, find the gallons of water that needs
     to be added to the bed.

                                                            _____

4.   What mechanisms do plants use to obtain water from soil?

Water Budgets

Water is vital for agriculture. In California a system has been devised to assist growers plan the efficient use of agriculture water. CIMIS (California Irrigation Management Information System) stations have been established in 85 locations around the state to provide weather information to growers and irrigation schedulers. All stations can be accessed by computer to monitor water use by crops for a growing area. If the soil's capacity to store water and the rate of use is known from CIMIS data then efficient irrigation scheduling may be achieved. Remember water use for an area greatly depends upon air temperature, wind speed, and relative humidity. Soil type has minimum influence. A sand must be irrigated more frequently than a clay loam to keep moisture conditions optimal for plant growth.

A. Go to the following URL (Uniform Resource Locator) on the Internet.

    http://www.dla.water.ca.gov/cgi-bin/cimis/cimis/data/input_form.pl

B. First, prepare two graphs by plotting, from **January to December**, the monthly precipitation and Eto (Evapo-transpiration) data for the Hopland FS (Field Station) in Mendocino County California and the same data from a station near your home town or county (Time on the "X" axis and "inches" of water on the "Y" axis). The map at the bottom of the web page can help you find the nearest home station. If you are not from California use the data from the San Luis Obispo Station.

1.   On the graphs shade the area where Eto is greater than **Precipitation** then determine the total water deficiency, in inches, for the year at the sites you have chosen?

2.   Determine the estimated water requirement* in acre-feet for the production of a crop at each location if the growing season was from April through August each year?

      Location _____   Location _____

      Acre-Feet _____   Acre -Feet _____

(*Estimated water requirement   = Σ(Eto - PRECIP) for each month during the growing season. Assume soil water storage and recharge are zero). Σ = summation.

3.   What three factors cause the estimated water requirement to be different for these two locations.

      1. _____ 2. _____ 3. _____

4.   Attach graphs, your calculations and answers to this page.

Instructor Copy   Name_____

                                                   Section _____

11.   Routine Particle-sise Analysis Results:

       Soil Sample _____

       A.   Grams of dry soil in the Bouyoucos cylinder.   _____ g

       B.   Hydrometer reading after 45 seconds.   _____ g

       C.   Grams of silt and clay in suspension.   _____ g

       D.   Grams of sand in the sample.   _____ g

       E.   Grams of clay in the sample.   _____ g

       F.   Percent sand, silt, and clay in the sample.   _____._____._____

       F.   Soil Texture Class   _____

Questions on Soil Texture:

1.   How would the water temperature affect the falling velocity of small soil particles in
       a suspension? (Hint: which factors in Stokes Law are affected by temperature)

2.   Find the time required for large clay (0.002 mm) particles to settle to the bottom of a
       three meter deep pond using Stokes Law.

3.   What effect could the humus left in the 50 gram soil sample above have on determining
       the soil's texture?

Instructor Copy   Name_____

Section _____

4.   The clay content of soils affects several soil properties. On the generalized graphs
        below, place the letter corrosponding to each soil property on the left side of graph
        A or B that best illustrates the relationship of the soil property to increasing clay
        content in soil.

            a. swelling potential   b. soil drainage ease   c. nutrient holding ability
            d. structure formation   e. ease of compaction   f. % humus
            g. ease of tilling   h. landscaping top soil   i. dry mass per cubic foot

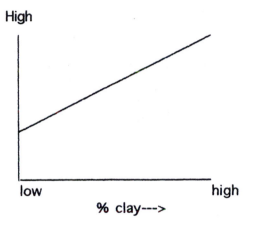

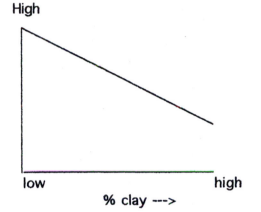

LABORATORY EXERCISE V

BULK DENSITY, POROSITY, AND PARTICLE DENSITY OF SOIL

Goal: To understand the influence of soil texture and structure on soil bulk density, particle density and porosity.

Objectives:

1. Understand the definition of all words used in this exercise.

2. Measure and calculate bulk density, particle density and porosity of soils that have different textures and structures.

3. Know the range of bulk density, particle density, and porosity values for most soils.

4. Know why the average particle density of mineral soils is 2.60 to 2.65 g/cm^3.

5. Use the bulk density and particle density measurements to calculate soil mass, volume, and porosity.

LABORATORY EXERCISE V

BULK DENSITY, POROSITY, PARTICLE DENSITY OF SOIL

(See Section 3:5, pages 78-81 in the Text)

Soil is a unique three-phase system composed of solids (soil particles), liquid (water), and gases (the soil atmosphere). The characteristics of, and the interactions among, these three phases determine the soil's physical properties. Bulk density, particle density and porosity are three soil physical properties that best illustrate the relationships among the soil phases. For example, the volume of voids in soil is related to the percentages of sand, silt and clay (i.e., the soil texture), and to the arrangement of these soil particles into peds (i.e., the soil structure).

The texture and structure of soil determine the size of the pores and the total porosity of the soil. This pore space in soil is important for root growth, water retention, atmospheric gas exchange, and water drainage. About fifty percent of the total volume of an ideal soil for plant growth will consist of pore space. Sands contain less pore space than any of the other textures, and clay usually has the most. The porosity of a soil can be reduced by compaction or increased by the addition of organic matter to improve a soil's structure. Soil porosity can be measured directly using water or calculated from the soil's bulk density and particle density.

Bulk density is the dry mass of a soil divided by its volume. The bulk density value for soil is expressed as follows:

$$\text{Bulk Density} = \frac{\text{oven dry soil mass (g)}}{\text{total soil volume (cm}^3)}$$

Some typical values for the bulk density of soils having different textures along with the corresponding porosity percentages are given in Table 10.

Table 10. Typical values of bulk density and porosity for soils

Soil Texture	Bulk Density Range	Average	Porosity Range	Average percent
	Grams per cubic cm.			
Sands and Sandy loams	1.2 to 1.8	1.6	30 to 55	40
Loams and silt loams	1.1 to 1.6	1.3	40 to 60	50
Clay loams and Clays	1.0 to 1.5	1.2	45 to 65	55

For a particular soil, the bulk density may vary due to compaction or loosening as a result of tillage operations. This means that the porosity will also change. Some soils, such as the surface horizons of forest soils, Histosols, and soils derived from volcanic ash, have very low bulk densities (sometimes less than 1.0 gram per cubic centimeter).

The **Particle** density of a soil is its dry mass divided by the volume of soil particles.

$$\text{Particle Density} = \frac{\text{Oven dry soil mass (g)}}{\text{Volume of soil particles (cm}^3)}$$

Particle density takes into account the mass and volume occupied by the solid particles only. It excludes the volume occupied by air and water. Since a large portion of most soils is composed of particles derived from minerals that contain 70 percent or more silica and oxygen, the particle density of most soils is approximately 2.65 grams per cubic centimeter. This particle density is nearly the same as the density of quartz. Variations in the particle density of soils are due to the presence of heavier minerals (such as iron oxides) or lighter organic matter. Kaolinite and illite clays have particle densities of about 2.65 g/cm^3 while montmorillonite clays have particle densities of about 2.4 g/cm^3 . Although tillage results in a change in both bulk density and porosity, it does not affect particle density. The particle density remains constant because tillage and other short- term changes do not alter the total amount or the chemical composition of the soil mineral particles.

Porosity, or percent pore space, is the volumetric percentage of the soil that is occupied by pores.

$$\text{\% Porosity} = \frac{\text{Volume of soil pores (cm}^3)}{\text{Total volume of the soil (cm}^3)}$$

Porosity is expressed as a percent (i.e. pph). The total volume of pores is usually greater in a well-structured fine-textured (clayey) soil than in a coarse-textured (sandy) soil. Pores tend to be larger in sandy soil than in clayey soils.

If the bulk density and the particle density of a soil are known, the porosity can be found by using the following relationships.

% porosity (Ep) = volume pore (Vp)/volume solid (Vs) + volume pores (Vp)

total soil volume (Vt), 100% = % solid volume (Vs) + % pore volume (Vp)

% Porosity (Ep) = 100% - (bulk density/particle density X 100)
Note: (BD/PD X 100) = % solid

% Porosity (Ep) = (100% - (BD/PD X 100)

Although a soil's total porosity is important, its pore size distribution is equally important. Individual pores can be categorized generally as macropores, which are larger than micropores. The large (macro) pores drain freely and allow free movement of air and water. These macropores promote soil aeration (free movement of gases) and water infiltration and drainage. Soils containing a large proportion of macropores are usually very sandy and tend to retain little water. The small (micro) pores retain more water, drain slowly and have restricted air and water movement. Restricted soil aeration causes reduced plant growth because roots need oxygen. Many microorganisms also require oxygen. Some biological and chemical reactions are inhibited by poor aeration.

Aggregation (granulation) or the clustering of the soil particles into aggregates creates larger macropores (cracks) between the peds. Within each aggregate are smaller micropores for water retention. A balance between macropores and micropores is desirable for the production of most plants. Roots must have access to both air and water.

In this exercise the bulk density, particle density and porosity will be determined for two soils, one fine-textured and the other coarse-textured. These determinations will be made on disturbed (loosened and sieved) soil samples. In addition, five field soil cores are provided for bulk density and porosity determinations. These soil cores are used to illustrate the differences in soil physical properties caused by disturbed and undisturbed samples as well as for different soil textures. The volume and weight data are provided for each of these soil cores. The bulk density and porosity are to be calculated and the results recorded on the Data Sheet.

Bulk Density and Porosity of Soil Cores

Five soil cores in metal cylinders are provided. The volume of the cylinders and the mass of dry soil in each core are given. Using these data determine the bulk density of each sample. Assuming the particle density of the soil in all cores to be 2.65 g/cm^3, calculate the porosity (% pore space) for each sample. Record the data and include the soil texture and soil condition (disturbed, compacted, or undisturbed) of each soil sample.

Data on Soils Collected in Cores

Core number	Soil Texture and structure condition	Mass of dry soil - grams	Bulk Density g/cm^3	Porosity %
1				
2				
3				
4				
5				

Volume of soil cores = _____ cm^3

Questions about cores

1. How would the addition of organic matter to these soils affect their bulk densities?

2. Why are the % porosity values for the clay textured soils higher than those with a sandy texture?

Bulk Density Determination

Procedure:

1. Weigh a clean, dry 100 ml graduated cylinder, record its weight and fill it to the 100 ml mark with one of the assigned soils. Compact the soil by holding the cylinder in one hand and tapping the bottom lightly on the palm of your other hand. Add more soil to bring the volume to 100 cm^3. (Note that 1 ml = 1 cm^3.)

2. Weigh the cylinder plus soil and record the weight in the table below.

3. Transfer the soil to a dry 250 ml beaker.

4. Level the soil surface and set the beaker aside for the porosity determination.

5. Repeat steps 2 and 3 of the above procedure with the other assigned soil.

6. Calculate the bulk density of each soil and record results below. Be sure to show correct units and complete setup for each determination.

Table for Bulk Density Data:

Soil textural class:	_____	_____
Mass of cylinder + soil (grams)	_____	_____
Mass of empty cylinder (grams)	_____	_____
Mass of dry soil (grams)	_____	_____
Total volume of soil (cm^3)	_____	_____
Bulk density (g/cm^3)	_____	_____

Bulk Density Questions:

1. How do your values for bulk density compare to those given in Table 10 for soils of comparable texture?

2. From the porosity formula, what must the bulk density of a soil be if its porosity is 37% and a particle density of 2.65g/cm^3?

3. What management practice could be used to change the BD of a soil from 1.76g/cm^3 to 1.24g/cm^3?

4. What mass, in pounds, would the soil in the top six inches of an acre-furrow-slice weigh if it has a BD value of 1.47g/cm^3?

5. What mass, in kilograms, would the soil in the top 15 cm of a hectare-furrow-slice weigh if it has a BD value of 1.34g/cm^3?

Direct Measurement of Soil Porosity

1. Fill a 100 ml graduated cylinder to the 100 ml mark with tap water. Tilt the beaker containing the soil saved from step 3 page 105 to about a 45 degree angle. Pour the water very slowly onto the soil in the beaker. Stop adding water when the soil pores are filled (**saturated**) and the soil surface glistens.

2. Allow the soil to stand for several minutes to make sure the water has filled the very small soil pores. Any excess water can be returned to the graduated cylinder. Don't stir or shake the mixture.

3. Determine the amount of water used to exactly fill the pore space in the 100 cm^3 of soil. Subtract the reading of the water level on the side of the graduated cylinder from the original volume of 100 ml. Note that the water which was poured onto the soil exactly filled the pore space and thus equals the total pore volume by direct measurement.

4. Record your results below and make the necessary calculations. Be sure to show formula and complete setup and units.

5. Repeat steps 1 through 4 with the other soils.

Porosity Data: Soil _____   _____

Initial volume of water in cylinder (ml)   _____   _____

Volume of water remaining in cylinder ml)   _____   _____

Volume of water filling soil pores (ml)   _____   _____

Total volume of soil (cm^3)   _____   _____

Porosity percent from direct measurement   _____   _____

Porosity Questions:

1. Which soil will usually have the higher porosity, a sandy soil or a clayey soil? Why?

2. Find the mass in kilograms of a cubic meter of undisturbed soil that has a bulk density of 1.24 g/cm^3

_____ kg

3. How would the compaction of a soil with a BD value of 1.3 g/cm^3 to a BD value of 1.8 g/cm^3 affect water flow into the soil and plant growth? Explain.

4. Approximately what proportions of the pore space should be occupied by air and water when a soil is at an optimum condition for plant growth? Explain.

Particle Density Determination

1. Place 50.0 ml of tap water into a 100 ml graduated cylinder.

2. Carefully weigh a 50.0 gram sample of dry soil. Slowly pour all of the soil into the water in the graduated cylinder.

3. Cover the mouth of the graduated cylinder with the palm of the hand and invert the cylinder back and forth to mix the soil and water thoroughly. Check to see that there are no air bubbles near the bottom of the cylinder.

4. Add exactly 10 ml of water to a 25 ml graduated cylinder. Carefully rinse the sides of the cylinder containing the soil with the 10 ml of water. Determine the new water level beneath the floating organic material.

Note: The level of the soil in the graduated cylinder has no bearing on this determination.

5. Record your results below and make the necessary calculations and specify the units, formulas and setup.

6. Repeat the above procedure with the other soil.

7. Return all equipment and clean your bench area.

Particle Density Data: Soil textural class _____   _____

Mass of dry soil (grams)   _____   _____

Total volume of soil plus water (ml)   _____   _____

Volume of total water in cylinder   _____   _____

Volume of soil particles (cm^3)   _____   _____

Particle density (g/cm^3)   _____   _____

Particle Density Questions:

1. Does repeated soil cultivation change the particle density of soil?   Explain.

2. Why is the particle density of mineral soils always near 2.65g/cm^3?

3. How do the above soil porosity values obtained by direct measurement compare to the calculated values you obtain using the BD and PD values ? Explain.

    (Note: show all four values and discuss why they are the same or different)

LABORATORY EXERCISE V

Instructor copy

Name _____

Section _____

Data on Soils Collected in Cores

Core number	Soil Texture and structure condition	Mass of dry soil - grams	Bulk Density g/cm³	Porosity %
1				
2				
3				
4				
5				

Volume of soil cores = _____ cm³

Table for Bulk Density Data:

_____

Soil Textural class: _____   _____

Mass of cylinder + soil (grams)   _____   _____

Mass of empty cylinder (grams)   _____   _____

Mass of dry soil (grams)   _____   _____

Total volume of soil (cm³)   _____   _____

Bulk density (g/cm³)   _____   _____

Bulk Density Questions:

1. How do your values for bulk density compare to those given in Table 1 for soils of comparable texture?

2. From the porosity formula, what must the bulk density of a soil be if its porosity is 37% ?

3. What management practice could be used to change the BD of a soil from 1.76g/cm³ to 1.24g/cm³?

4. What mass, in pounds, would the soil in the top six inches of an acre-furrow-slice weigh if it has a BD value of 1.47g/cm³?

Name _____

Section _____

5. What mass, in kilograms, would the soil in the top 15 cm of a hectare-furrow-slice weigh if it has a BD value of 1.34g/cm^3?

Porosity Data: Soil textural class: _____   _____

Initial volume of water in cylinder (ml)	_____	_____
Volume of water remaining in cylinder ml)	_____	_____
Volume of water filling soil pores (ml)	_____	_____
Total volume of soil (cm^3)	_____	_____
Porosity percent from direct measurement	_____	_____

Porosity Questions:

1. Which soil will usually have the higher porosity, a sandy soil or a clayey soil? Why?

2. Find the mass in kilograms of a cubic meter of undisturbed soil having a bulk density of 1.24 g/cm^3

_____ kg

3. How would the compaction of a soil with a BD value of 1.3 g/cm^3 to a BD value of 1.8 g/cm^3 affect both water flow into the soil and plant growth? Explain.

4. Approximately what proportions of the pore space should be occupied by air and water when a soil is at an optimum condition for plant growth? Explain.

Instructor copy Name: _____

Section _____

Particle Density Data: Soil textural class _____ _____

Mass of dry soil (grams) _____ _____

Total volume of soil plus water (ml) _____ _____

Volume of total water in cylinder _____ _____

Volume of soil particles (cm^3) _____ _____

Particle density (g/cm^3) _____ _____

Particle Density Questions:

1. Does repeated soil cultivation change the particle density of soil? Explain.

2. Why is the particle density of mineral soils always near 2.65g/cm^3?

3. How does the above soil porosity values obtained by direct measurement using water compare to the calculated values you obtain using the BD and PD values determined in this lab? Explain.

   (Note: show all four values and discuss why they are the same or different)

<u>Porosity Summary</u>

	Direct Measurements	Calculated Porosity
Sandy Texture	_____ %	_____ %
Clay Texture	_____ %	_____ %

LABORATORY EXERCISE VI

FIELD STUDY OF LOCAL SOILS

Goal: To examine and describe two local soils developed on different land forms.

Objectives:

1. Understand the definition of all words used in this exercise.

2. Know the five soil forming factors and how they have affected the formation of local soils.

3. Determine the land form on which each soil profile formed.

4. Learn the properties for each of the following horizons A, E, B, C, and R.

5. Know how to determine the following soil properties in the field: texture, structure, color, pH, slope, water runoff class, and permeability.

6. Learn the major properties of a mollic and argillic diagnostic horizons.

7. Find and interpret soil series information on the internet.

Soil Profiles and Horizons

A **soilprofile** is a vertical cross-section of the soil through all its horizons and extending into the parent material. A soil **horizon** is a layer of soil approximately parallel to the land surface and differing from adjacent layers in physical, chemical, and biological properties. The following drawing, Diagram A, of a landscape depicts the relationship of soils to different parent materials. Every soil series is unique and recognizable. Each has developed as a product of the five soil forming factors, parent material, climate, biota, topography, and time. Soils are dynamic in their development. They slowly evolve over time through horizon development in a parent material as shown on the Diagram B below.

Diagram A. Cross Section of a landscape illustrating soil and parent material relationships.

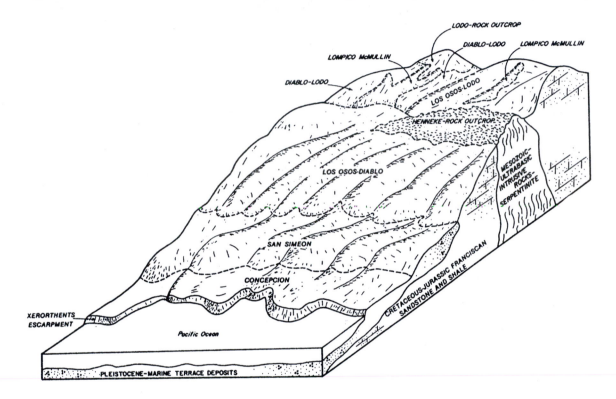

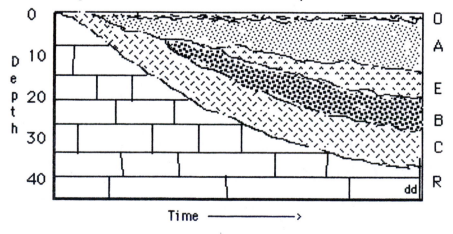

Diagram B. Master Soil Horizon Development from Bedrock

Soil horizons are typically described using the following format:

Horizon designation; depth; color (dry); texture; color (moist); structure; tilth; soil reaction (pH); boundry

Go to the following URL on the NET and find the official description of two different soil series. You may choose soil names from the lab display or you may try using your home town or local community name to find a soil. Soils are named after geographic locations where they were first found.

http://www.statlab.iastate.edu/cgi-bin/osd/osdname.cgi

Complete the following table for each soil:

Series Name   _____   Series name   _____

Location   _____   _____

Elevation   _____   _____

Soil Order   _____   _____

Parent Material:

_____   _____

Annual Rainfall _____   _____

Frost Free Days _____   _____

Permeability   _____   _____

USE   _____   _____

Vegetation   _____   _____

Distribution   _____   _____

            Epipedon present   Endopedons present

     _____   _____

                                          _____

1.   Which soil series is older? Explain.

2.   What is the major limiting factor restricting the use of each of the above soils for intensive agriculture?

Research soil scientist have accumulated vast amounts of information about soils across the United States and the world. That data are often compiled into generalized interpretative tables that can be used to catagorize soils. Table 11 illustrates the relationships of slope and texture to runoff rates. The information serves as a guide to land use planning.

Table 11. Surface Horizon Texture, Slopes and Water Runoff Rates

Slope	Sands, Loamy Sands	Sandy Loams, Sandy Clay Loam, Clay Loam, Silt Loam, Silty Clay Loam, Loam	Silty Clay, Clay, Sandy Clay
(0-1%)	very slow	very slow	very slow
(1-2%)	very slow	slow	slow
(2-6%)	slow	medium	medium
(6-12%)	medium	rapid	very rapid
(12-18%)	rapid	very rapid	very rapid
(>18%)	very rapid	very rapid	very rapid

Permeability:

Soil permeability is that quality of a soil that enables it to transmit air or water. Texture, structure, cracking, and organic matter content influence permeability. The permeability of the least permeable horizon controls water and air flow in a soil. The soil permeability classes are given below.

Slow - Less than 0.6 inches per hour. Slow permeability includes textures of silty clay, clay and sandy clay and soils with massive subsoils.

Moderate - 0.6-6.00 inches per hour. Moderate permeability includes textures of silt loam, loam, sandy clay loam, silty clay loam and sandy loam.

Rapid - 6.00-20.00 inches per hour. Rapid permeability includes textures of loamy sand and sand and soils with greater than 15% gravel.

Soil Classification:

Soils in the United States are classified according to the USDA Soil Taxonomy System

Soil Taxonomy explains the system of soil classification published by the Soil Survey staff of the U.S. Department of Agriculture. This system is an attempt at a comprehensive natural classification of soils based on measurable and observable soil morphological, physical, and chemical properties.

Soil Taxonomy is the result of many years of observation. Probably the greatest utility of the taxonomy is for soil management and land use planning. The nomenclature is designed to fit into any modern language. Terms were coined mainly from Greek and Latin roots, and are used as mnemonic devices for remembering the names. Many soil properties can be described by the use of the formative elements. This system of classification is hierarchical with 11 orders subdivided into 54 Suborders, 238 Great Groups, 1922 Subgroups, more than 5,000 Families, and more than 18,000 Series.

The **Orders**, the highest level of the classification, are distinguished on the basis of properties that mark soil forming processes on the grand scale. A soil order is based upon the presence of distinctive pedogenic features. Most taxa have features known as diagnostic surface and subsurface horizons. (See Section 7:2 pages 219-226 and the Plates between pages 160-161 in the Text)

A. **The diagnostic surface horizons** are called **epipedons**, from the Greek words epi over), and pedon (soil). The epipedon includes the upper part of the soil darkened by organic matter, the upper alluvial horizons, or both. Six epipedons are commonly recognized. Two epipedon descriptions follow. (See Section 2:8, pages 46-52 in the Text)

1. **Mollic epipedon:** A thick (>7 inches), dark (darker value and chroma than 3.5 when moist) surface horizon with a high percent (>50%) base saturation, high in calcium, good structure and not hard and massive when dry. The horizon contains at least 1% organic matter.

2. **Ochric epipedon:** A horizon that is too light, too low in organic matter, too thin to be mollic, or is hard and massive when dry.

B. **The diagnostic subsurface horizons** (endopedons) characterize different soils in the system. There are several subsurface horizons described in the U.S. Soil Taxonomy. Not all soils have a diagnostic subsurface horizon.

1. **Argillic horizon.** This is a horizon of illuviation (accumulation) of clay. If the soil is not eroded, this horizon is in the position of the B horizon and it has more clay than the overlying A or E horizon. The Argillic horizon is generally equivalent to the Bt horizon. Field clues for identifying Argillic horizons are:

   (I) Higher in clay than horizon above and generally higher than horizon below.
   (II) Horizon will have soil structure, often blocky or prismatic.
   (III) Clay films are sometimes visible to the naked eye or with hand lens. There are smooth, shiny clay coatings on the soil peds and in pore indicating illuviation.

2. **Cambic horizon:** This is an altered or changed horizon in the position of a B horizon. This horizon has been changed by internal physical movement or by chemical reactions to such an extent that it no longer retains the original nature of the rocks or sediments in the C horizon. Some chemical changes have occurred but generally some minerals that could be weathered rather easily are still present. Some leaching has usually taken place and color changes have occurred, but there is no evidence of any clay movement or illuviation (i.e. clay films). The cambic horizon is generally equivalent to a Bw horizon.

There are eleven orders in Soil Taxonomy. A brief definition of three orders follows.

1. **Entisol:** These are the "young" soils with no diagnostic horizons. They may be formed in fresh alluvium on floodplains, in fresh sand dune deposits, or on steep, eroded hillslopes where diagnostic horizons either cannot form or have not had time to form.

2. **Mollisol:** These are the soils of the native grasslands characterized by having a mollic epipedon and high base saturation. They may also have an argillic horizon but do not exhibit evidence of a high shrink-swell potential such as deep (>50 cm) surface cracks in the dry season.

3. **Vertisol:** These are the soils that contain greater than 30% montmorillonite clay. There is evidence of a high shrink- swell potential such as deep (>50 cm), wide (>1 cm) surface cracks in the dry season and subsurface slickenside (shiny ped surfaces) production in the wet season.

The Soil Forming Factors
(See Section 2:4, pages 28-36 in the Text)

A. Soil Parent Material
The material soil develops from is called parent material. When rocks are exposed to atmospheric conditions, they begin to adjust to their new environment. This adjustment, known as weathering, involves processes which cause physical disintegration and chemical decomposition of the rocks. The weathering of bedrock produces unconsolidated debris that serves as the parent material for soils. The parent materials undergo continued alteration and evolve into a soil that reflects the integrated effects of climate, biotic factors (plants, animals and microorganisms), topography and time.

The following parent materials are common.

1. **Alluvium:** Unconsolidated material transported and deposited by flowing water.

2. **Colluvium:** Unconsolidated material deposited on and at the base of steep slopes by direct gravitational action.

3. **Residuum:** Unconsolidated weathered mineral matter that accumulates by disintegration and decomposition of bedrock in place equivalent to a Cr horizon.

B. Topography or Landscape Position
Soils occur in a predictable mosaic pattern on the earth's surface. All soils are located on a specific percent slope in a specific landscape position. These are called **landforms.** (See Figure 3 below).

1. **Flood Plain** (FP): The nearly level fluvial plain that borders a stream or river and is subject to inundation and sediment accumulation under floodstage conditions. It is a constructional landform built of sediment deposited during overflow and lateral migration of the stream.

2. **Stream Terrace** (ST): A natural level strip of land in a stream valley, flanking and more or less parallel to the stream channel, originally formed near the level of the stream, and representing the dissected remnants of an abandoned floodplain, stream bed, or seashore.

3. **Toeslope** (T): The landform component that forms the outermost, gently inclined surface at the base of a hillslope. Toeslopes are constructional surfaces forming the outermost point of a hillslope where alluvium tends to accumulate.

4. **Footslope** (F): The landform component that forms the inner, gently inclined surface at the base of a hillslope. The surface is dominantly concave in profile. It is a transition zone between the backslope and toeslope where colluvium and alluvium tend to accumulate.

5. **Backslope** (B): The landform component that forms the steepest inclined surface of the hillslope. Backslope in profile are commonly steep and linear, or steep and convex-concave. Backslopes are erosional forms produced mainly by mass wasting (direct gravitational action) and running water action.

6. **Shoulder** (SH): The landform component that forms the uppermost inclined surface at the top of a hillslope. It comprises the transition zone from backslope to summit of an upland. The surface is dominantly convex in profile and erosional in origin.

7. **Summit** (SU): The top or highest level of an upland feature such as a hill or mountain.

Fig. 3 Idealized Landform Cross Section

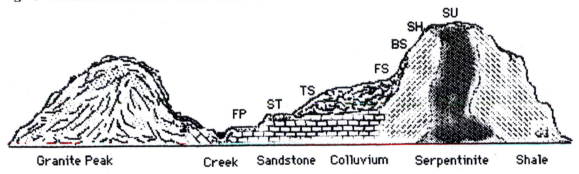

Granite Peak   Creek   Sandstone Colluvium   Serpentinite   Shale

C. Climate

Climate is the soil forming factor that helps determine soil differences on a regional scale (i.e. arid desert vs. humid forest) or on a local level (i.e., micro climatic differences). The important components of climate that influence soil formation include precipitation, and temperature.

As precipitation increases from one locale to another, there is more water available to move downward through the soil profile. As water percolates downward (this process is called leaching), water- soluble soil constituents such as soluble salts (bicarbonates, chlorides, nitrates) are translocated from upper soil horizons and are moved deeper in the soil profile. Other materials including clay minerals do not dissolve in water but are suspended in the soil solution and may leach into subsoil horizons. Therefore, the greater the amount of precipitation, the deeper these dissolved and suspended materials are leached. The consequences of leaching include a lower pH (higher acidity), fewer basic cations (Ca, Mg, K, Na), and a lower soil fertility. As a result, there will be little leaching in desert soils and extensive leaching in humid forest soils.

Soil temperature regulates the amount of evapotranspiration. The higher the temperature, the greater the rate of evapotranspiration. In hot, dry regions, any precipitation added to the soil may quickly evaporate into the atmosphere. However, in cool, moist regions, precipitation accumulates in the soil and leaching results. If two regions have the same amount of precipitation, the region having the lower soil temperatures will have more "effective" precipitation and soils will be more highly leached in the cooler region. On a local scale in the Northern Hemisphere, this concept explains the differences between soils on north-facing slopes with those on south-facing slopes. The soils on north-facing slopes receive less direct solar radiation (lower temperatures, less evaporation). These soils are, therefore, more highly weathered, deeper, and subject to greater leaching.

121

D.   Time

Once the solid rock materials were weathered or the alluvial and colluvial materials and aeolian dunes were deposited, plants began to grow. Nitrogen was added to soil from rainfall and the nitrogen-fixing activity of lichens, algae, and legumes. Soil organic matter began to accumulate and the soils developed a distinct dark colored A horizon. Rainwater began to dissolve and leach the calcium and magnesium carbonates. Some soils were so high in lime that leaching has not depleted the supply and these soils remain calcareous. Generally the lime inhibits further soil profile development by reducing downward leaching of clay particles.

On the soils formed from dunes and non-calcareous sandstone and shale materials the leaching process has removed some of the exchangeable basic cations (Ca, Mg, K, Na) resulting in a slightly acid soil (pH 6 to 7). On the more permeable materials the rainwater has had sufficient time to translocate silicate clay minerals into the B horizon. The clay accumulation filled the pore spaces and decreased the permeability of the B horizon (Los Osos soils). This zone of illuvial clay accumulation is defined as an argillic horizon.

In poorly drained soils, little leaching occurs. These soils have suffered from a lack of oxygen resulting in reduction of iron. When this process has occurred periodically the soil exhibits mottles, while flooded soils exhibit olive and gray colors indicative of gleying.

Because of the extensive erosional processes that have occurred the youngest soils may overlie the oldest geologic formations. Young soils have developed on stabilized sand dunes, on the flood plains and on the steep mountain uplands. Some of the oldest soils have developed on alluvial terraces and alluvial fans of the Paso Robles Formation.

An example of soil profile development as a function of soil age for a soil formed in unconsolidated parent material under tall grass vegetation is shown in Fig.4. Notice how soil profile development and productivity vary with time.

E.   Living Organisms

The activity of living organisms has a major influence on the development of soils. Burrowing animals, including ants, gophers, worms, and prairie dogs, mix the surface layer and tend to cause fewer but deeper horizons to form. Clay and organic matter movement downward is interrupted. Soils developed under humid forest differ greatly from those formed under grassland. They often have many horizons and are highly leached. In contrast soils formed under grassland have fewer horizons and often contain unleached lime and nutrients. There is a marked accumulation of humus in the "A" horizon. The microbial decompositon of organic matter and humus produces acids that enhances the release of nutrients from soil minerals. Cultivated prairies around the world produce abundant harvests of wheat, corn, and soybeans.

On the Field trip today your instructor will assist you with the completing the following field data sheets used to collect information about soils.

MAP SYMBOL

SOIL SERIES. TYPE. PHASE

SURFACE COARSE FRAGMENTS (> 2M MI%)

	GRAVEL	COBBLE	STONE	BOULDER

DATE | **BY** | **PHOTO NO** | **STOP NO**

AREA | FOREST | RANGER DISTRICT | STATE | COUNTY | LOCATION SEC | T | R

POTENTIAL NATURAL VEGETATION

POTENTIAL VEGETATION

LANDFORM | PARENT ROCK AND FORMATION

SLOPE % | SLOPE LENGTH (FT) | SINGLE | COMPLEX | ASPECT | ELEVATION FEET METERS | MEAN ANNUAL-PRECIP INCHES CM | MEAN ANNUAL AIR TEMP °F | SOIL TEMP AT 50 CM °C

HORIZON	DEPTH IN CM	COLOR — PED DRY MOIST	COLOR — CRUSHED DRY MOIST	MOTTLING	TEXTURE	STRUC-TURE	CONSIST-ENCE DRY MOIST WET PLAST	CLAY FILMS	GRAVEL COBBLE STONE % VOLUME	ROOTS	PORES	pH	CARBONATE (EFF)	BOUND-ARY	DIAGNOS-TIC HORIZONS

CLASSIFICATION

PEDON DESCRIPTION

USDA – FOREST SERVICE

| MAP SYMBOL | SOIL SERIES, TYPE, PHASE | | | | |

| AREA | | FOREST | RANGER DISTRICT | STATE | COUNTY | DATE | BY | PHOTO NO | STOP NO |

POTENTIAL NATURAL VEGETATION

POTENTIAL VEGETATION

| SLOPE % | SLOPE LENGTH (FT) | SINGLE | COMPLEX | ASPECT | ELEVATION |

SURFACE COARSE FRAGMENTS (> 2M MI%

| GRAVEL | COBBLE | STONE | BOULDER |

LOCATION

SEC   T   R

PARENT ROCK AND FORMATION

LANDFORM

| MEAN ANNUAL PRECIP | MEAN ANNUAL AIR TEMP | SOIL TEMP AT 50 CM |
| INCHES   CM | °F   °C | |

ELEVATION — FEET — METERS

| DEPTH | COLOR | | | TEXTURE | STRUC-TURE | CONSIST-ENCE | CLAY FILMS | GRAVEL COBBLE STONE | ROOTS | PORES | pH | CARBON-ATE (EFF) | BOUND-ARY | DIAGNOS-TIC HORIZONS |
| IN / CM | PED DRY MOIST | CRUSHED DRY MOIST | MOTT-LING | | | DRY MOIST WET PLAST | | % VOLUME | | | | | | |

HORIZON

CLASSIFICATION

125

Laboratory Exercise VI
Homework

Instructor Copy

Name _____

Section _____

Go to the following URL on the NET and find the official description of two different soil series. You may choose soil names from the lab display or you may try using your home town or local community name to find a soil. Soils are named after geographic locations where they were first found.

http://www.statlab.iastate.edu/cgi-bin/osd/osdname.cgi

Complete the following table for each soil:

Series Name   _____   Series name   _____

Location   _____   _____

Elevation   _____   _____

Soil Order   _____   _____

Parent Material:

_____   _____

Annual Rainfall _____   _____

Frost Free Days _____   _____

Permeability   _____   _____

USE   _____   _____

Vegetation   _____   _____

Distribution   _____   _____

Epipedon present   Endopedons present

_____   _____

                               _____

1.   Which soil series is older? Explain.

2.   What is the major limiting factor restricting the use of each of the above soils for intensive agriculture?

LABORATORY EXERCISE V

SOIL FORMATION QUESTIONS

Name_____

Section _____

1. Name two components of climate that greatly influence soil formation.

    _____   _____

2. List five types of soil parent materials found in our local area.

    _____   _____   _____   _____

    _____

3. How would soil temperature and evaporation differ for soils occurring on north versus south facing slopes?

4. How would the activity of gophers affect the soil formation process?

5. Find the Soil Series Description for the profiles examined on the field trip using the URL above. Compare what you observed in the field with the official description. How are the soils different from the official description? Attach your answer to the back of this page.

6. How does soil texture change with depth for each soil profile studied ? Explain.

7. How does soil pH change with depth for each profile studied? Explain.

8. In the space below draw three graphs that indicates the changes that occur in soil with time.

Organic matter content		clay content in sub- -soil		Soil Fertility	
	Time		Time		Time

LABORATORY EXERCISE VII

SOIL ORGANIC MATTER, HUMUS AND MICROBIAL ACTIVITY

Goal: To understand the role of soil microorganisms in cycling organic matter and nutrients.

Objectives:

1.   Understand the definitions of all words used in this exercise.

2.   Measure microbial transformations of carbon and nitrogen in a soil.

3.   Understand how the carbon to nitrogen ratios (C/N) of organic soil amendments influence N availability to plants.

4.   Measure the amount of humus in soil by chemical extraction.

5.   Learn the nitrogen and carbon cycles.

LABORATORY EXERCISE VII

SOIL ORGANIC MATTER, HUMUS AND MICROBIAL ACTIVITY

The solid portion of soil is composed of minerals and organic matter. The organic matter includes plant and animal residues at various stages of decomposition, cells and tissues of soil organisms, and substances synthesized by the soil biota. Each of these contains many types of compounds including proteins, sugars, polysaccharides such as cellulose, hemicellulose, and starch, and complex fats, waxes, and lignins. All of the life essential elements are contained in the collection of organic compounds found in soils.

The total amount of organic matter in soil and the nutrients it contains vary with climate, nature of the parent material, soil pH and the kind and amount of vegetation produced on or applied to the soil. The process of soil organic matter decomposition and humus formation may be represented as a partial oxidation process.

organic matter + O_2 + microorganisms ---> CO_2 + H_2O + Energy (heat) +

humus + new microbial cells + plant nutrients (N,P,S,etc.)

Under natural soil conditions organic residues undergo an initial rapid decomposition that converts them to inorganic and simplified organic products. In the process carbon dioxide, water, energy and nutrients in the organic materials are released. One of the most important functions soil micro-organisms perform is to recycle organic carbon to carbon dioxide. The carbon dioxide released is used by plants in the photosynthesis process to build more plant material for microbial decomposition.

The process of burning organic residue releases nutrients to the soil also. Nitrogen and sulfur however, are converted to gases and are lost to the atmosphere by this rapid oxidation process. The principal reasons for adding organic residues to soils are (1) to modify the tilth of the soil, making seed bed preparation easier, (2) to add plant nutrients, and (3) to dispose of unusable or unwanted organic waste. Other benefits of adding organic matter to soils include an improved soil structure, increased cation exchange capacity, increased water holding capacity, increased aeration, and a reduction in soil erosion.

Humus is the amorphous, heterogeneous fraction of the soil organic matter remaining after the major portion of added plant and animal residues have decomposed. Usually it is dark brown to black in color. Generally, it is bound to the surfaces of clays, rendering it resistant to microbial attack. Humus has a low bulk density (0.05 to 0.25 grams per cubic centimeter), is very finely divided (colloidal), and is an important reservoir of essential elements including N, Ca, Mg, P, S, Cu, Zn, and Mn. The humus content of arable soils averages between 2-3 % and ranges from less than 1 % up to 12 %.

Mineralization

Mineralization is the process of converting elements from an organic form to an inorganic form, usually by microbial decomposition. After the addition of organic materials to soil, the rate of release of carbon dioxide reaches a peak in a few days and then tapers off with time. The peak is associated with an explosive growth in the microbial population. They compete vigorously with plants for nutrients needed for building new cells. The decline in carbon dioxide production occurs because the microbes have consumed most of the readily decomposable high energy organic materials. Carbon dioxide production and microbial population is dramatically reduced simultaneously. As the microbial activity declines, the dead cells decompose, releasing nutrients in a plant available inorganic form. Nitrogen is released as ammonium (NH_4^+). Sulfur is released in a reduced form, possibly as hydrogen sulfide (H_2S). Most other nutrients, including phosphorus, are released in ionic form (e.g. $H_2PO_4^-$). The mineralization of large quantities of nitrate and sulfate signal the completion of the major microbial degradation process.

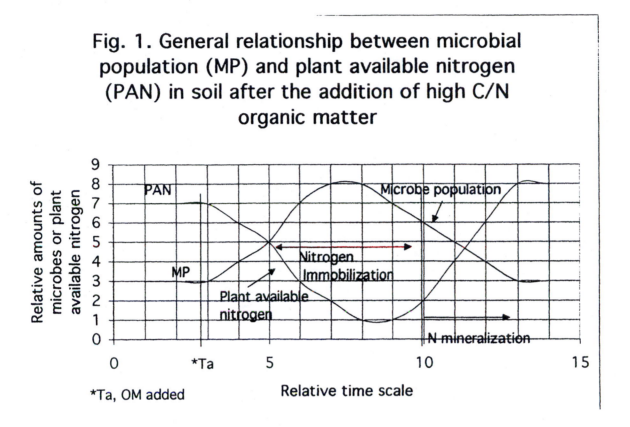

Fig. 1. General relationship between microbial population (MP) and plant available nitrogen (PAN) in soil after the addition of high C/N organic matter

Soil Micro-organisms

Soil organisms usually comprise less than 0.5% of the dry soil mass. The population of living organisms in one heaped tablespoon of topsoil is generally greater than the number of people on earth. There is little of a practical nature one can do to alter permanently the microbial population of a soil. The soil normally contains such a vast array and abundance of organisms that most direct additions of organisms to this community will be ineffective. Usually, microbial populations will be altered more by changing the crop growing on the soil than by adding other microorganisms. This is not necessarily true, however, in soils that have been steam sterilized , or fumigated to control pathogens. The average biomass distribution found in an acre-furrow-slice of soil is shown in Table 12. (See pages 178-179 in the Text)

Table 12. Source and estimated distribution of biological materials in soil.

Source	Biomass* kg/hectare -15 cm	Number per AFS[+] or per gram of soil
A. Living		
earthworms	200 - 500	1.5×10^5/ AFS
insects	100 - 200	5.0×10^3/AFS
rodents	50 - 100	1.30×10^2/AFS
nematodes	10 - 100	10 - 50/gram
plant roots	3800	
bacteria	400 - 5000	2.6×10^9/gram
actinomycetes	300 - 4000	1.4×10^8/gram
fungi	1000 - 10000	3.7×10^6/gram
protista	15 - 200	1.3×10^4/gram
algae	50 - 600	2.8×10^5/gram

B.   Non-living (including humus, 6496 pounds/AFS or 5800 kg/HFS)

* Dry weight estimate based upon three percent organic matter in soil
+ AFS equals an acre-furrow-slice, approximately 2×10^6 pounds soil.

A number of conditions affect microbial populations in soils. The optimum temperature range for decay organisms is between 20 and 40°C (about 70 to100°F). Soil temperatures outside this range will retard the activity of most soil organisms. Excessive water in soil reduces the numbers and kinds of living organisms due to poor aeration. However, at low moisture levels soil micro-organisms thrive better than higher plants. Fungi, bacteria, and actinomycetes numbers vary with soil pH. If the pH of the soil is less than 6.0, fungi become the dominant soil microorganisms. The supply of nutrients, organic material for energy, and free oxygen gas also affect microbe numbers. Generally, optimum soil conditions for plants and soil microorganisms are similar.

C/N Ratio

(See text pages 200 - 202)

The carbon-organic nitrogen ratio (C/N) is the ratio of the mass of organic carbon to the mass of organic nitrogen in soil, organic material, or microbial cells. The proportion of carbon to nitrogen in the organic matter is an important factor influencing soil microbial activity. Organic material having a high carbon to nitrogen ratio (C/N), such as wheat straw at about 80/1, will decay slowly because the material contains insufficient nitrogen to satisfy the growth requirements of the decay causing microbes. Plant available soil nitrogen quickly becomes utilized by soil microorganisms for their growth and unavailable to higher plants until after the microbes expire. Nitrogen **immobilization** by microbes can create nitrogen deficiencies in the soil and lead to reduced plant growth. Legume residues, such as clovers and alfalfa, have low C/N ratios (< 30/1) and decay very rapidly in the soil. A list of common organic materials and their C/N ratios is shown on the following page in Table 13.

Table 13. The C/N ratio for some common organic materials*

Material	% carbon	% nitrogen	C/N ratio
Microorganisms	50	6.2	8/1
Humus, mollisol	56	4.9	11/1
Alfalfa, young	40	3.0	13/1
Clover hay	40	2.0	20/1
Barnyard manures	35	1.4	25/1
Sewage sludge	48	1.7	28/1
Rye grass	40	1.1	37/1
Peat moss	48	0.8	60/1
Corn stalks	44	0.6	65/1
Barley straw	45	0.1	80/1
Hardwood sawdust	46	0.1	400/1
Spruce sawdust	50	0.05	600/1

* Data compiled from numerous sources. See Table 6.6 in the text.

The Carbon Cycle

Microbial decay of organic matter releases copious amounts of carbon dioxide. The carbon dioxide escapes from the soil into the atmosphere. Plants complete the carbon cycle by absorbing CO_2 from the atmosphere for use in photosynthesis to produce more organic matter for decay. The whole process is often thought of as "the cycle of life".

Green plants ----> organic material -----> decomposition ----> CO_2 + Water

plus energy   ----> photosynthesis in green plants ----> more organic material.

The Nitrogen Cycle

The nitrogen in humus and other organic materials is mineralized by microbes during the decay process and can further undergo several transformations in the soil. For example ammonium (NH_4^+) is oxidized first to nitrite (NO_2^-) by **Nitrosomonas** and then to nitrate (NO_3^- ions) by **Nitrobacter** microbes. Ammonium and nitrate ions can be immobilized by plants and soil microorganisms or undergo other changes as illustrated in the nitrogen cycle (Figure 2).

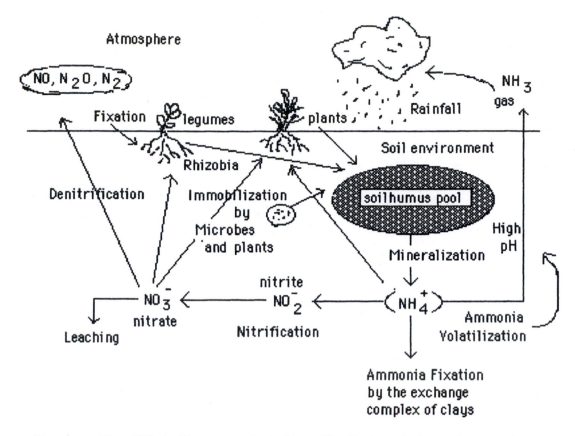

Figure 2. Simplified nitrogen cycle in the soil, plant, and atmosphere system

Definitions:

1. Fixation -- the conversion of nitrogen gas(N_2) to ammonia and subsequently to organic nitrogen utilizable in biological processes.
2. Mineralization -- the conversion of organic nitrogen to inorganic nitrogen by microbes.
3. Nitrification -- the oxidation of ammonium ions to nitrate ions
4. Immobilization -- the assimilation of ammonium and nitrate into tissue
5. Denitrification -- the reduction of nitrate to nitrogen gases in anaerobic soils
6. Volatilization -- the chemical conversion of ammonium ions to ammonia gas in high pH soils
7. Leaching-- the loss of nitrate to ground water below the root zone
8. Ammonia Fixation -- adsorption of ammonia ions (NH_4^+) into interlayer positions of the soil clay mineral fraction that cannot be replaced by neutral 1.0 N KCl.

Incubation Procedure:

1. Place 1 rounded tablespoon of moistened soil (free of stones and trash) into each of three clean plastic incubation bottles. The Instructor will ensure the soil has the proper moisture content.

2. Label the bottles with tape as "control", "low C/N amended" and high C/N amended. Write your name and lab time on the tape for each bottle.

3. Add 0.5 gram of pulverized low C/N organic material to the bottle labeled "low C/N amended" and mix thoroughly with the soil.

4. Add 0.5 grams of pulverized high C/N organic material to the bottle labeled "high C/N amended and mix thoroughly.

5. Do not add organic material to the "control" bottle.

6. Date experiment begins _____

   Low C/N Amendment _____   High C/N amendment _____

   % N _____ C/N _____   %N _____ C/N _____

7. Obtain 3 glass vials and fill each about 2/3 full with sodium hydroxide (NaOH) solution using the dispenser. (Caution NaOH is very caustic, wash after use)

8. Tilt the plastic bottle on its side to expose the bottom. Carefully place the glass vial with NaOH into the bottle. Do NOT spill the NaOH on the soil.

9. Carefully place the lid on each bottle and secure tightly.

10. Place the bottles aside to be incubated as directed by the instructor. (usually 1-2 weeks at room temperature)

11. Clean and put away all materials.

Analysis Procedure

1. After the specified incubation period, retrieve the 3 plastic bottles from the incubation area. Be careful. Do NOT spill the NaOH from the glass vials.

2. Open the bottles and remove the vials containing the NaOH. Label the vials with a small piece of tape to keep track of the treatments; C = control, L = low C/N amended, and H = high C/N amended. Place the vials in a test tube holder.

3. Examine the soil surface in each bottle for fungi. A microscope may be used to see differences among treatments. Fungal mycelia look like spider webs.

4. Rank the treatments in order of increasing numbers of mycelia observed.
   0 = none, 1 = few, 2 = many, 3 = large amounts

   Control _____   Low C/N _____   High C/N _____

Carbon Dioxide Test

1. Add 10 drops of Reagent 1 (barium chloride, $BaCl_2$) to each glass vial.

    Note-A white milky precipitate (barium carbonate $BaCO_3$) indicates the presence of carbon dioxide that was absorbed by the sodium hydroxide from organic matter decomposition by microbes during the incubation period.

2. Allow the precipitate to settle for about 5-10 minutes. Record on the data sheet the relative amounts of white precipitate in each vial and the date (0 = none, 1 = low, 2 = medium, 3 = high).

    Control _____   Low C/N _____   High C/N _____

    Date   _____ Days of incubation _____

3. Rinse the contents of the test vials into the designated barium waste container for recycling. (Caution, barium is a toxic heavy metal and is not biologically safe to pour down the drain or ingest, wash your hands).

QUESTIONS:

1. Was there a difference in the amount of CO_2 produced by each treatment? Explain.

2. Draw a simplified diagram of the carbon cycle tha includes plants and microbes. See the carbon cycle description above.

3. Why doesn't the earth's atmosphere contain huge amounts of carbon dioxide as a result of microbiological activity?

Nitrogen Extraction Procedure

1. Add 25 ml of ammonium acetate solution to the soil in each bottle.

2. Place the cover on each bottle and mix by shaking for about one minute, then let the samples stand about 2 minutes to settle.

3. Arrange three filter funnels lined with Whatman No. 1 filter paper on a filtration rack. Label and place a beaker under each funnel to catch about 10 ml of filtrate.

4. Pour only the extracting solution liquid from the plastic bottles into the appropriate filter funnel and collect 10 ml of filtrate from each bottle.

5. During the filtration process clean the plastic bottles and vials. Return each to the storage containers.

Nitrate Test

1. Place a small spoonful (0.1g) of powdered **Reagent** 2 into each beaker containing the nitrogen extracts.

2. Swirl each beaker for about one minute and allow the suspension to settle.

3. Fill a clean small test vial 1/2 full with the clear nitrogen extract by decanting. Label each vial appropriately.

4. Add 10 drops of **Reagent** 3 to the liquid in each test vial. Mix with a clean glass stirring rod.

   A reddish color indicates the presence of nitrate in the extract. (the more reddish the color the more nitrates present).

5. Record the relative amounts of nitrate in each vial here and on the data page, pp 143.
   (0 = none, 1 = low, 2 = medium, 3 = high)

   Control _____   Low C/N amended _____   High C/N amended _____

Questions:

1. Compare the amount of available nitrogen in the control to each of the amended samples. Indicate which of the amended samples best demonstrate the process of immobilization and mineralization of nitrogen?

2. Explain why there is a difference in available nitrogen between the amended samples.

3. What practical steps can be taken to reduce nitrogen immobilization by soil microorganisms when turning into the soil for disposal, high C/N rice straw after the grain harvest?

Humus Test

The humus content of soil can be estimated from a sodium pyrophosphate extract of the soil. Humus is made soluble and leached from the soil with a 0.1N sodium pyrophosphate-0.1N sodium hydroxide solution. The dispersed humus can be collected and used to estimate the soil humus content.

1. Place 5 grams of your desk soil onto the center of a single piece of folded filter paper fitted into a funnel on a filtering rack. Place a 250 ml beaker under the funnel to catch the filtrate.

2. Add 30 ml of the 0.1 N sodium pyrophosphate-0.1 N sodium hydroxide solution to the soil and collect the filtrate. This is the humus extract.

3. Fill a test tube 3/4 full with the humus extract and estimate the humus content of the soil from the color of the extract as indicated below:
   (see the humus extract standards provided by the instructor)

   Table 13. Sodium pyrophosphate-sodium hydroxide extract color and estimated humus content of soils

Color of Extract	Humus Level	Humus %
very light brown	very low	<1
light brown	low	2
brown	medium	3
very dark brown	high	4
black	very high	5

Color of your extract _____ Humus content estimate _____%

4. Clean all equipment and return it to its proper place, and then clean your bench area.

5. Answer the following questions and complete the data summary page.

   a. Under what natural conditions would the humus content of soils increase?

   b. What activities of mankind contribute to the loss of humus from soil?

   c. Give five reasons why humus is important to the soil environment.

140

6. Use the estimated percent of humus in your desk soil and the average composition of humus to calculate the pounds of humus, carbon, and nitrogen in an acre furrow slice (2 million pounds) of soil.

a. percent humus _____

c. pounds carbon _____

b. pounds humus _____

d. pounds of total nitrogen _____

Instructor copy

Name _____

Section _____

Soil Microbial Activity

4. Rank the treatments in order of increasing numbers of mycelia observed.
    0 = none, 1 = few, 2 = many, 3 = large amounts

    Control ____   Low C/N ____   High C/N _____

    Rank the treatments in order of the amount of CO_2 produced.
    (0 = none, 1 = low, 2 = medium, 3 = high).

    Control ____   Low C/N ____   High C/N _____

    Date   _____ Days of incubation _____

5. Record the relative amounts of nitrate in each vial here.
    (0 = none, 1 = low, 2 = medium, 3 = high)

    Control ____   Low C/N amended ____   High C/N amended _____

Questions:

1. Was there a difference in the amount of CO_2 produced by each treatment? Explain.

2. Draw a simplified diagram of the carbon cycle. See the carbon cycle description above.

3. Why doesn't the earth's atmosphere contain huge amounts of carbon dioxide as a result of microbiological activity?

LABORATORY EXERCISE VII

Instructor copy

Name _____

Section _____

Questions:
1. Compare the amount of available nitrogen in the control to each of the amended samples. Indicate which of the amended samples best demonstrate the process of immobilization and mineralization of nitrogen?

2. Explain why there is a difference in available nitrogen between the amended samples.

3. What practical step can be taken to reduce nitrogen immobilization by soil microorganisms when turning into the soil for disposal, high C/N rice straw?

Humus Test: Color of your extract _____ Humus content estimate _____ %

5. Answer the following questions

   a. Under what natural conditions would the humus content of soils increase?

   b. What human activities contribute to the loss of humus from soil?

   C. Give five reasons why humus is important to the soil environment.

      1.

      2.

      3.

      4.

      5.

6. Use the estimated percent of humus in your desk soil and the average composition of humus to calculate the pounds of humus, carbon, and nitrogen in an acre furrow slice (two million pounds) of soil.

   a. percent humus _____   c. pounds carbon   _____

   b. pounds humus _____   d. pounds of total nitrogen _____

Instructor Copy

Name _____

Section _____

1. Will soils high in organic matter and humus always have a high level of micro-biological activity ? Explain your answer.

2. What is LISA and how might it help us produce more food?

3. What symbiotic and non-symbiotic organisms are involved in the transfer of nitrogen from the air to soil, to humus and ultimately to other living organisms?

4. What are the six most important factors that affect the decomposition of organic waste and humus in soil(See text, Miller and Gardiner PP 200-203).

   A.   B.   C.

   D.   E.   F.

5. What are the advantages and disadvantages of composting organic waste materials?

   Advantages   Disadvantages

   1.   1.

   2.   2.

   3.   3.

LABORATORY EXERCISE VIII

SOIL TESTING AND FERTILIZERS

Goal: To learn the fundamentals of soil testing programs, to know some basic terminology used in fertilizer management, and learn to do fertilizer calculations.

Objectives:

1. Understand the definition of all words used in this exercise.

2. Know the five key requirements of an effective soil testing program.

3. Determine the amounts of calcium removed from two soil by a nutrient extracting solution.

4. Understand the basic concept of correlation and its value to soil testing.

5. Be able to Identify the following fertilizer terms: carrier, filler, grade, complete fertilizer, amendment, agricultural mineral.

6. Know how ammonium fertilizers can be lost from soil by volatilization.

7. Be able to calculate the elemental percentages, weights and costs of nutrients in fertilizers when given the grade.

LABORATORY EXERCISE VIII

SOIL TESTING AND FERTILIZERS
(See Section 12:4, pages 383-385 in the Text)

The goal of soil testing is to measure the fertility status of soil, and to provide the grower a fertilizer recommendation that will ensure optimum crop yield. Soil testing is practiced in the United States and throughout the world with varying degrees of success. In order for a soil testing program to be successful, the following five requirements must be met:

1. A sample representative of field conditions

    Soil tests are usually done on 10 to 50 grams of soil that have been taken from a composite of samples collected in the field. The composite sample must be air-dried and crushed small enough to pass through a 2 mm sieve. Special care must be taken to avoid conditions in both the field and the laboratory that would make the samples non-representative. For accuracy and reliability, laboratory managers prefer to have their own trained technicians collect samples for testing.

2. A well equipped laboratory.

    The basic equipment and instrumentation required for standard soil tests may cost up to $250,000 at 1996 prices. The following are also important for the smooth operation of a soil testing service:

    A. technicians skilled at doing chemical analyses,
    B. a support staff to do clerical work,
    C. field representatives to collect samples.

3. A proper nutrient extracting solution

    The purpose of the nutrient extracting solution is to simulate nutrient removal by plant roots. Various chemical solutions are used to remove nutrients from a soil sample. These solutions can be acids, salts, organic compounds or water, depending upon which plant nutrient is to be evaluated. Extracting solutions vary by regions throughout the country because soils vary. An extracting solution that works well on soils from one area will not always work well for soils from another location. The best nutrient extracting solution to use is the one that most closely approximates plants' ability to obtain a given nutrient from soil.

4. A method of correlating the results of a soil test with an expected plant response

The selection of nutrient extracting solutions used for soil tests requires both field and laboratory research. Carefully planned studies are required to find the best nutrient extracting solution to use for determining if a soil contains sufficient plant nutrients for high yields. From Figure 1, notice how the yield of corn varies with the amount of phosphorus removed from a soil by an extracting solution. In this case, the yield of corn is highly correlated with the amount of phosphorus the solution extracts from the soil. Soil testing becomes meaningless without the correlation data to relate soil test values to expected plant growth.

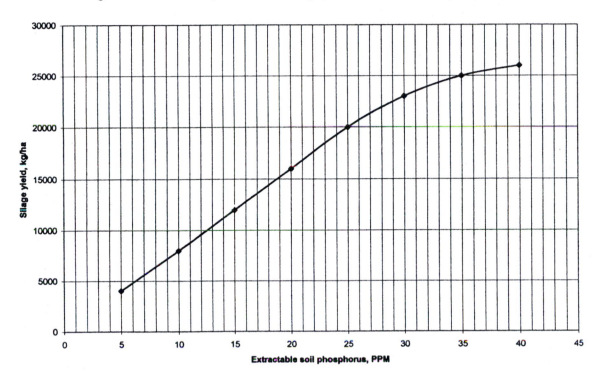

Fig.1. General relationship between corn silage yield and extractable soil phosphorus

5. A proper interpretation of the soil test data

The ultimate reasons for soil testing are (1), to Identify which plant nutrients in a soil are in short supply, and (2), to determine the amount of each nutrient to be added to a growing area for optimum crop production. The grower often relies on an agronomist for advice when selecting the most economical application rate of fertilizer for crop production.

Even if all five requirements are met, soil testing cannot eliminate crop yield loss from weeds, disease, unfavorable weather, improper irrigation, or poor management.

Part 1. Calcium Extracting Solution

Calcium is an essential plant and animal nutrient. It is a vital part of plant cell walls and animal bones. The calcium in soils can be (1) a component of a mineral, (2) held as an exchangeable cation on clay particles (3), a part of the humus fraction, (4) a free cation in the soil solution. Plant roots are capable of obtaining some calcium from all of these sources. The best we can do with soil testing is to find methods that correlate well with plant growth response and allows us to predict if the soil contains sufficient plant food. The following procedure illustrates the principle involved in nutrient extraction, correlation, and interpretation.

1. Fit two funnels with filter paper and place them on the filter rack. Place a clean beaker under each funnel.

2. a. Place 10 grams of your desk soil into a flask that have been labeled D.
   b. Place 10 grams of a sandy loam soil into a flask that has been labeled S.

3. Add 50 ml of 0.1 N HCl to each flask.

4. Swirl the flasks for one minute and filter.

5. Test for calcium in each filtrate as follows: in separate clean test tubes place 10 ml of each filtrate. Add 10 drops of sodium hydroxide (NaOH) and 5 drops of ammonium oxalate to each test tube. If calcium is present a white precipitate of calcium oxalate will form. Mix the solutions thoroughly by covering the test tubes with a small piece of parafilm and shaking.

6. The amount of calcium extracted from each soil can be estimated by comparing the amount cloudness of your extracts with instructor prepared standards of 10, 20, and 40 ppm calcium using a standard nephelometer or basic colorimeter. (A 1:1 dilution with pure water may be required). Use a 10 ml graduated cylinder for dilution, five mls extract to five mls of pure water.

   Transfer the solution in step 5 or the diluted solution in 6 to a special test tube that fits into the nephelomether or basic colorimeter. (Instructor needs to prepare a Ca standard curve with 10, 20, 40, and 100 PPM Ca).

7. Results:

A. Estimated parts per million calcium in: B. Estimated pounds Calcium per acre foot soil in:

   Desk sample _____ppm   Desk sample _____ pounds/acre-foot

   Sandy loam _____ ppm   Sandy loam _____ pounds/acre-foot

Note: an acre-foot of soil weighs about four million pounds.

Questions:

1. Turn to Table 13-1 in the text, pp415. List the crops likely to need additional calcium added to the soil if grown in the Desk soil and the Sandy loam. (assumes perfect correlation)

   Desk soil   Sandy loam

2. A soil test for available potassium shows the soil contains 105 ppm K_2O. Accord to Table 13.1, page 415 in the text, is there enough of the nutrient in an acre-furrow-slice of soil for growing 400 bushels of potatoes?

3. The result of a soil test for plant available phosphorus was 26 ppm. If a grower obtained a corn yield of 625 kg/ha, would this mean that the soil test was not valid? Explain. (Refer to Fig. 1, page 112).

PART II. Fertilizer Materials   (See Section 12:8, pages 403-411 in the Text)

Commercial   fertilizers are a combination of chemicals that contain available plant essential elements. When marketed, a fertilizer must contain a total of $N + P_2O_5 + K_2O > 5\%$. Any material containing less than 5% of the plant food elements must be labeled as an **agriculture mineral**. Any material sold for the purpose of improving soil physical conditions for plant growth is called an **amendment**. There are two major types of fertilizers: (A) inorganic, and (B) organic (see Table 14). A display of several fertilizers is provided for you to use to complete this part of the laboratory.

Fertilizer grades such as 0-45-0 and 11-48-0 are called **incomplete   fertilizers**. A **complete   fertilizer** is a blend of materials that contain the three major fertilizer elements.

Any substance that contains a plant nutrient is considered a **carrier**, i.e., super phosphate, $Ca(H_2PO_4)_2$ is a carrier of phosphorus. When mixing carrier materials to obtain a special grade of fertilizer such as 5-10-15, **fillers** (lime, sand, clay, or gypsum) are included to complete the weight requirement for a given quantity of fertilizer of a chosen grade. The weight basis is usually a ton.

Note that each fertilizer has three numbers on its label to identify its **grade**. The first number is the % total N, the second the % available P_2O_5 , and the third the % water soluble K_2O in the fertilizer. The actual amounts of phosphorus (P) and potassium (K) in a fertilizer are found by determining the amount of P in P_2O_5 and the amount of K in K_2O using atomic weights. To find the actual amount of phosphorus in P_2O_5, first, find the gram atomic weight of P_2O_5 as follows:

2 phosphorus (P)	@ 31g	=	62g
5 oxygen (0)	@ 16g	=	80
			142 grams equals the weight of P_2O_5

The atomic weight ratio of $2P/P_2O_5$ (62/142) equals 0.44. This ratio can be used to determine the elemental amount of phosphorus in a fertilizer with the following formulas:

Eq. A   0.44 X pounds P_2O_5 = pounds P in a fertilizer

Similarly for K_2O,

2 potassium (K)	@ 39g =	78g
1 oxygen (0)	@ 16g =	16g
		94 grams equals the weight of K_2O

The atomic weight ratio $2K/K_2O$ (78/94) equals 0.83

Eq. B   0.83 X pounds K_2O = pounds K in a fertilizer

Example calculation
    A fertilizer has a grade of 8-16-12. Find the actual amount of N, P, and K in 200 pounds of the material.

0.08 X 200 =   16 pounds N,   0.16 X 200 X 0.44 = 14 pounds P
    0.12 X 200 X 0.83 =   19.9 pounds K

Table 14. Common Fertilizer Materials

Name	Chemical Formula	Form	Grade
A. Inorganic Fertilizers			
Ammonium nitrate	NH_4NO_3	Solid	34-0-0
Monoammonium phosphate	$NH_4H_2PO_4$	Solid	11-48-0
Diammonium phosphate	$(NH_4)_2HPO_4$	Solid	18-46-0
Ammonium phosphate sulfate	$(NH_4)_3H_2PO_4SO_4$	Solid	16-20-0
Anhydrous ammonia	NH_3	Gas	82-0-0
Aqua ammonia	NH_3H_2O	Liquid	20-0-0
Calcium nitrate	$Ca(NO_3)_2$	Solid	15.5-0-0
Urea	$(NH_2)_2CO$	Solid	46-0-0
Urea ammonium nitrate	$(NH_2)_2CO\ NH_4NO_3$	Solid	32-0-0
Concentrated superphosphate	$Ca(H_2PO_4)_2$	Solid	0-45-0
Potassium nitrate	KNO_3	Solid	13-0-44
Potassium sulfate	K_2SO_4	Solid	0-0-52
Potassium chloride	KCl	Solid	0-0-60
B. Organic Materials			
Steer manure			2.0-0.5-1.9
Sheep manure			2.0-1.0-2.5
Poultry droppings			4.0-3.2-1.9
Seaweed (kelp)			0.2-0.1-0.6
Fish meal (dry)			10.4-5.9-0
Sewage sludge			2.0-3.0-0
Bone meal			1.0-12.0-0
Dried Blood			12-1.5-0
Tankage			7.0-8.6-1.5
Bat Guano			13-5-2

*Organic materials are highly variable in moisture content, and in the concentration of individual plant nutrients.

PROCEDURE:

Ia. Select a complete fertilizer from the organic group and another from the inorganic group of samples on display or from Table 14 and answer the following questions:

ORGANIC FERTILIZER A

Sample Name _____

Grade _____

INORGANIC FERTILIZER B

Sample Name _____

Grade _____

Ib. How many pounds of nutrients would a ton of each fertilizer contain? Show complete setup of problem using dimensional analysis.

FERTILIZER A

Pounds N _____

Pounds P_2O_5 _____

Pounds K_2O _____

FERTILIZER B

Pounds N _____

Pounds P_2O_5 _____

Pounds K_2O _____

Ic. What Is the actual amount of N, P, and K in 500 pounds of each fertilizer?

FERTILIZER A

Pounds N _____

Pounds P _____

Pounds K _____

FERTILIZER B

Pounds N _____

Pounds P _____

Pounds K _____

2. If a grower needs to apply 300 pounds of nitrogen to a field, how many pounds of each of the following materials must be applied to meet the nitrogen requirement? Show complete setup using dimensional analysis.

POUNDS

_____Fertilizer A

_____Fertilizer B

_____Urea

_____Diammonium phosphate

_____Poultry droppings

3. What is the grade of a pure $MgNH_4PO_4$ fertilizer material?
(Mg = 24, N = 14, P = 31, O = 16, H = 1)

Part III.   Ammonia Volatilization

The application of ammonium (NH_4^+) forms of nitrogen to soils that have a pH greater than 7.0 will often result in the loss of ammonia by volatilization. The NH_4^+ ions react with the OH^- ions in the soil solution and are changed to NH_3 gas molecules. Studies have shown that from 10 to 60% of the ammonium nitrogen applied to alkaline soils can be lost by ammonia volatilization in a short period of time. The use of other forms of nitrogen, injecting the fertilizer into the soil and the acidification of soil are techniques used to minimize the loss of nitrogen by ammonia volatilization.

$$NH_4^+ \quad + \quad OH^- \text{--------->} \quad NH_3 \quad + \quad H_2O$$

(ammonium)   (hydroxide)   (ammonia gas)   (water)

PROCEDURE

1.   Mark two clean 125 ml flasks A and B.

2.   Place about 10 granules of diammonium phosphate fertilizer in each flask. Dissolve the fertilizer with about 10 ml of deionized water.

3.   To flask A add 10 drops of acid (1 N HCl) and mix. Test for the presence of ammonia gas by holding a strip of wet litmus paper inside the flask without touching the sides of the flask. If ammonia (NH_3) Is being volatilized the paper will turn blue. Ammonia gas leaving the flask also can be detected by smell.

4.   To flask B add 10 drops of base (1 N NaOH) and mix. Repeat the tests for ammonia as before.

RESULTS: Which flask is losing ammonia gas?

Flask A _____   Flask B _____
acidic (pH < 7)   basic (pH > 7)

QUESTION:

1.   Explain the results of your volatilization experiment.

2.   What practical steps can a grower take to reduce the loss of applied nitrogen fertilizer by ammonia volatilization from alkaline soils?

3.   Under what pH condition should animal waste water be kept to reduce the loss of ammonia gas?

FERTILIZER HOMEWORK

1. If an ounce of 12-8-5 fertilizer for house plants costs $1.98, determine the cost of purchasing one pound of plant food when using this fertilizer.

                                               _____

2. If ammonium nitrate fertilizer costs $246.00 per ton, find the cost of the fertilizer in cents per pound. (NH_4NO_3 = 34-0-0)

                                               _____

3. Find the cost of elemental nitrogen, in cents per pound, in the ammonium nitrate fertilizer above.

                                               _____

4. What would it cost to fertilize 4500 square feet of lawn with ammonium nitrate if the recommended application rate for lawn grass is 320 pounds of fertilizer per acre?

                                               _____

5. If urea (46-0-0) cost $320.00 per ton and ammonium sulfate (21-0-0) cost $287.00 per ton, which fertilizer would be the most economical source of nitrogen? Explain.

                                               _____

6. What are three serious consequences of over application of fertilizers?

LABORATORY EXERCISE VIII

INSTRUCTOR COPY

Name _____

Section _____

EXTRACTING SOLUTION RESULT:

A. Estimated parts per million calcium in: B. Estimated pounds Calcium per acre foot soil in:

Desk sample _____ppm   Desk sample _____ pounds/acre-foot

Sandy loam _____ ppm   Sandy loam _____ pounds/acre-foot

Note: an acre-foot of soil weighs about four million pounds.

Questions:

1. Turn to Table 13-1 in the text, pp415. List the crops likely to need additional calcium added to the soil if grown in the Desk soil and the Sandy loam.

   Desk soil   Sandy loam

2. A soil test for available potassium shows the soil contains 105 ppm K_2O. Accord to Table 13.1, page 415 in the text, is there enough of the nutrient in an AFS of soil for growing 400 bushels of potatoes?

3. The result of a soil test for plant available phosphorus was 26 ppm. If a grower obtained a corn yield of 625 kg/ha, could this mean that the soil test was not valid? Explain. (Refer to Fig. 1, page 112).

LABORATORY EXERCISE VIII

Instructor copy

Name _____

Section _____

Ia. Select a complete fertilizer from the organic group and another from the inorganic group of samples on display or from Table 14 and answer the following questions.

ORGANIC FERTILIZER A   INORGANIC FERTILIZER B

Sample Name   _____   Sample Name   _____

Grade   _____   Grade   _____

Ib. How many pounds of nutrients would a ton of each fertilizer contain? Show complete setup of problem using dimensional analysis.

FERTILIZER A   FERTILIZER B

Pounds N   _____   Pounds N   _____

Pounds P_2O_5 _____   Pounds P_2O_5 _____

Pounds K_2O _____   Pounds K_2O _____

Ic. What Is the actual amount of N, P, and K in 500 pounds of each fertilizer?

FERTILIZER A   FERTILIZER B

Pounds N   _____   Pounds N   _____

Pounds P   _____   Pounds P   _____

Pounds K   _____   Pounds K   _____

2. If a grower needs to apply 300 pounds of nitrogen to a field, how many pounds of each of the following materials must be applied to meet the nitrogen requirement? Show complete setup using dimensional analysis.

POUNDS

_____Fertilizer A

_____Fertilizer B

_____Urea

_____Diammonium phosphate

_____Poultry droppings

3. What is the grade of a pure $MgNH_4PO_4$ fertilizer?
(Mg = 24, N = 14, P = 31, O = 16)

LABORATORY EXERCISE VIII

Instructor copy

Name _____

Section _____

AMMONIA VOLATILIZATION

RESULTS: Which flask is losing ammonia gas?

Flask A _____   Flask B _____
acidic (pH < 7)   basic (pH > 7)

QUESTION:

1.   Explain the results of your volatilization experiment.

2.   What practical steps can a grower take to reduce the loss of applied nitrogen fertilizer by ammonia volatilization from an alkaline soils?

3.   Under what pH condition should animal waste water be kept to reduce the loss of ammonia gas?

Fertilizer Costs:

1. If an ounce of 12-8-5 fertilizer for house plants costs $1.98, determine the cost of purchasing one pound of plant food when using this fertilizer.

_____

2. If ammonium nitrate fertilizer costs $246.00 per ton, find the cost of the fertilizer in cents per pound. (NH_4NO_3 = 34-0-0)

_____

3. Find the cost of elemental nitrogen in cents per pound in the ammonium nitrate fertilizer above

_____

4. What would it cost to fertilize 4500 square feet of lawn with ammonium nitrate if the recommended application rate for lawn grass is 320 pounds of fertilizer per acre?

_____

5. If urea (46-0-0) cost $320.00 per ton and ammonium sulfate (21-0-0) cost $287.00 per ton, which fertilizer would be the most economical source of nitrogen? Explain.

_____

6. What are three serious consequences of over fertilizing plants?

Soil Survey Report Laboratory IX
(See Section 20, pages 630-652 in the Text)

GOAL: This exercise has been designed to acquaint you with a Soil Survey Report.

Objectives:
1. to learn how to use the Legal Land Description for describing parcels of land.
2. to discover the broad range of useful information found in a Soil Survey report.
3. to know the difference between the Storie Index and Land Capability Class, LCC
4. to estimate range land productivity
5. to understand how to use soil information in making land use decisions.

Soil Surveys began in the United states in 1898. Their initial purpose was to study agriculture and forestry lands. Modern surveys have been expanded to include information for the multipurpose use of land. They include interpretative information on the use of soils for wildlife management, park and recreation development, road construction, building foundations, septic tank leach field suitability, fill material selection, forest management, production agriculture and waste management.

As the population of the United States expands it makes sense to have an inventory of our land resources. It is vital for the informed decision making process of land use. We have no more land but there will be many more people in the future seeking to use the limited resource. The Soil Survey is a key information source in the development of a stewardship plan to wisely use the soil so it will continue to support us and future generations.

Legal Land Description: George Washington's interest in surveying helped promote the adoption of the "rectangular survey system" by the Federal Government on April 26, 1785. Since then the General Land Office has surveyed most of the United States using the system. The Legal Land Description applies to 30 of the 50 states excluding the 13 original ones and the Mexican Land Grants.
See the map of California, Appendix C for an example of the Legal Land Description.   Beginning with principle meridians, square areas six miles on a side are established north-south to form Townships and east-west to form Ranges. There are three principle meridians in California, San Bernardino, Mount Diablo, and Humbolt. Each square, six miles on a side, is called a township. Each township is divided into 36 sections. Each section is one square mile and contains 640 acres. Each section can be subdivided into quarters of 160 acres. Every quarter section can be further subdivided into smaller parcels as shown on in Fig. 1 parcel "R". An example of a Legal Description follows: (See Appendix C).

See Appendix C

Practice Exercise

"R" example: NW 1/4, NE 1/4, Sec 9, T2N, R2E. Mount Diablo BLPM (40 Acres)
Find Parcel B: W1/2, SE1/4, SW1/4, Sec 9. T2N, R2E, Mount Diablo BLPM (20 Acres)

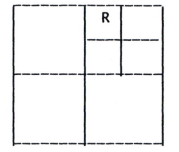

Soil Survey Report Laboratory IX

Instructor Copy

Name _____

Section _____

1. Complete the following table for soils (a - d) from information found in the interpretative tables of the SLO County Soil Survey Report, Coastal Part. See Map Sheet 9.
   a. (174) Mocho loam 0-2% slope
   b. (121) Concepcion loam 5-9% slope
   c. (160) Los Osos loam 15-30% slope
   d. (127) Cropley clay 0-2% slope

Table of Soils Interpretative Information

Map unit symbol	a. 174	b. 121	c. 160	d. 127
Soil Series	Mocho	Concepcion	Los Osos	Cropley
Texture of surface				
Slope %				
Depth to high water table (ft)				
Soil pH (surface)				
Depth to bedrock (in)				
Permeability (in/hr) (slowest horizon)				
Flooding Duration				
Shrink-swell potential __Highest				
Storie Index (0-100)				
Land Capability class (I - VIII) irrigated				

2.   Complete Table 1 to show the yields of row crops being produced on the four soils.

Table 15 - Normal Yields per Acre for Selected SLO County Crops

	Lemons	Vine Grapes	Lettuce	Celery	Broccoli	Garbanzo	Barley
Soil	Pounds	Tons	Crates	Pounds	Tons	Pounds	Pounds
174							
121							
160							
127							

(The regular price of lemons is $800/ton, grapes $975/ton, lettuce $20/cwt, celery $15/cwt, broccoli $30/cwt, garbanzo $44/cwt and barley $2.48/bushel (56 lbs). (A cwt is equal to 100 pounds.)

3.   Which individual soil and crop combination produces the highest gross income per acre?

Soil_____

Crop_____

Income/acre $_____

4.   Find the normal range land productivity for the above four soils. Why is there no rating for soil 174, Mocho?

121   _____lbs/ac

127   _____lbs/ac

160   _____lbs/ac

5.   If a brood cow consumes 25 pounds of dry food on rangeland per day, how many acres of each of these three soils would be required to safely support a herd of 50 cows for a year without supplemental feeding?

121 _____ acres

127 _____ acres

160 _____ acres

6.   Which of the above four soils is best suited for row crop agriculture? Explain your choice using the completed interpretative table, the Storie Index (pp111) and the Land Capability Classification rating(pp 109).

Answer

7. How are the above four soils being used by current landowners? (See the general soil descriptions)

8. Use a sheet of clear plastic and a wax pencil to trace an outline of the section of land on Map Sheet 9 containing each of the following geographic features. In the boxes below give the complete legal description for each location.

    a. Cuesta Park to the nearest 40 acres.

    _____

    b. The Cal Poly Environmental Horticulture Department green houses to the nearest 10 acres.

    _____

    c. Mission San Luis Obispo de Tolosa to the nearest 5 acres.

    _____

9. Which of the four soils, Mocho, Los Osos, Cropley, or Concepcion is best suited for the construction of a water storage pond? Explain.

    Answer_____

10. Which of the four soils is best suited for irrigation? Explain.

    Answer_____

11. What is the major limitation of each of the four soils for installing a septic tank leach field for the disposal of home waste water?

12.   Use Table 15 to determine the soil order for each of the four soils.

    174 _____121 _____127 _____ 160 _____

Laboratory Exercise X

Soils and the Environment

(See Ch. 19 pages 595-626 in the text)

GOAL: To conclude the Introductory Soil Science class with a walking field trip of campus to explore both ecological and environmental applications of Soil Science principles to the sustainability of agriculture and other land use operations.

Objectives:

1. Review and observe the application of soil science principles in the field.

2. Observe examples of Class I, Class II, Class IV, Class VIII, and Class VII land.

3. Final quiz on Soil Survey and field trip.

Laboratory Exercise XI

Salt Affected Soils

Goal: To develop in students a fundamental understanding of how sodium salts affect soil physical properties and plant growth.

Objectives:

1.   Know the definitions of key terms and units of measure used in this exercise.

2.   Understand the difference in effect sodium and calcium have on water infiltration into soil.

3.   To be able to calculate the amount of total dissolved solids(TDS) in water given the Electrical Conductivity.

4.   To measure the effect TDS can have on osmotic potential and plant available water.

5.   Develop and understanding of how a conductivity meter works.

6.   Know the common "salt sinks" used to dispose of unwanted salts in soils.

7.   Understand how sodic soils can be reclaimed for agriculture productivity.

Laboratory Exercise XI
Salt Affected Soils

Sec. 9.1-9.3 PP 284-294

In many parts of the world where irrigated agriculture is extensively practiced, salt accumulation in soil often limits the type of crops grown and requires the use of special management practices to keep the growing area productive. Installing drainage ditches, and tile lines below the soil surface provide a means of removing excess salt from soil. Water percolating through the soil and eventually into ditches or into the tile lines carries the soluble salt to an area for disposal. The common "salt sinks" used to rid a growing area of unwanted salt include rivers, lakes, evaporation ponds, the ocean, or in most cases the soil itself. The major source of salts in soil comes from applications of irrigation water. The most common ions found in irrigation water include, Na^+, HCO_3^-, Ca^{++}, Mg^{++}, Cl^-, SO_4^{--} and NO_3^-.

Drainage water may also contain many other environmentally important ions. Trace amounts of selenium found in some California irrigation waste water, for example, has been found to cause, through biomagnification up the food chain, reproductive abnormalities among waterfowl.

The amount of soluble salt in water and soil is commonly measured using a Conductivity Meter that records **Electrical Conductivity, (EC).** The more dissolved salt in water or a water extract of soil, the greater the EC reading. The common units used to express the value for EC are either millimho/cm (mmho/cm), micromho/cm (μmho/cm), or deciSiemens/meter, (dS/m). The numerical values for mmho/cm and dS/m are the same.

Three general chemical conditions usually develop in soils in arid climates. These condition include saline, sodic, and saline-sodic. The table below lists the criteria used to define each condition. Very salt sensitive plants are affected by EC values as low as 2.0 dS/m. (Strawberries, some clovers)

Table A. Soil salinity criteria for salty conditions found in soils.

	dS/m	ESP*	usual pH**	SAR
"Normal"	<4.0	<15%	6.5-7.5	<13
Saline	>4.0	<15%	6.5-7.5	<13
Sodic	<4.0	>15%	7.5-9.0	>13
Saline-Sodic	>4.0	>15%	>7.5	>13

* exchangeable sodium percentage, (see PP 291-293 in the text)
** not a diagnostic criteria, SAR = Sodium Adsorption Ratio, see pp 292.

Total dissolved solids (TDS) in soils and water can be estimated from EC measurements. For soil, a water extract at saturation is normally used to assess the salinity condition. Where the soil contains substantial amounts of expanding clays or peat it becomes difficult to obtain a saturation extract. A different soil/water ratio is used (1:1 - 1:5). The following empirical relationships are used to estimate the amount of total dissolved salts in soil and their effect on plant available water.

A. dS/m X 640 = TDS in mg/L or PPM

B. dS/m X 0.36 = Osmotic Potential of a solution in bars

C. dS/m X 36 = Osmotic Potential (OP) in kPa

If a sample of well water had a EC reading of 0.56 dS/m, it would contain about , (640 X 0.56), 358 mg/L of TDS (358 PPM TDS). The water's osmotic effect on plant roots would be minimal since the OP value is below field capacity, 1/3 bar or 33 kPa.

Procedure: Part I

1. Each student is assigned to bring to class a sample of water from home.

2. Measure the Electrical Conductivity (EC) of the following liquid samples and determine the TDS and Osmotic Potential,(OP) for each sample.
    (The instructor will demonstrate how to use the Conductivity Meter. To obtain an exact reading, the EC meter values needs to be adjusted for solution temperature and cell constant). Warm water conducts current better than cold water and EC cell design affects readings. (See appendix D)

	Adjusted Readings		
Sample	dS/m	TDS, mg/L	OP, kPa
a. Tap water	_____	_____	_____
b. Deionized or distilled water	_____	_____	_____
c. Sample from home	_____	_____	_____
d. Soda water	_____	_____	_____

3. Transfer 25 grams of two designated soils to a 250 ml conical flask and add 50 mls of DI or distilled water.

4. Stopper the flask and shake contents for 10 minutes. Transfer the liquid to a funnel fitted with Whatman #1 filter paper and collect about 10 mls of clear solution.

5.   Determine the EC, TDS, and OP for each soil. Since a 2:1 water to soil ratio was used, the EC value needs to be doubled to account for a dilution effect.

<u>Corrected EC</u>

Sample	dS/m	TDS, mg/L	OP,kPa
Soil A	_____	_____	_____
Soil B	_____	_____	_____

Answer the following questions.

1.   Find the quantity of salt in pounds added to an acre of land by the application of an acre-foot of irrigation waters a, b. and c. above. (An acre-foot of water is equal to 1.2×10^2 liters of water. Show calculations.

2.   For the soda water, find the amount of TDS in grams one ingests when consuming each 20 ounce soft drink.

_____ g

3.   Compare the soda water TDS amount with regular tap water TDS and explain which is better for quenching thirst. Explain.
(Note, potable drinking water must contain less than 500 mg/L TDS).

Tap water   _____mg/L

Soda water   _____ mg/L

180

Effect of Na and Ca on soil water percolation rates.

Procedure: Part II

1.   Place three small funnels on a filter rack. Insert Whatman #1 filter paper into each funnel and add 30 grams of a dry clay loam soil to each.

2.   Place a 50 ml graduated cylinder under each funnel to catch the leachate.

3.   Slowly add DI water to funnel one and record the time required for 15 ml of water to percolate through the soil in the table below.

4.   Slowly add 1.0 N $CaCl_2$ funnels to the soil in funnel two and record the time required to collect 15 ml of leachate in the table below.

5.   Add 1.0 N NaCl to the soil in funnel three and record the time required to collect 15 mls of this leachate in the table below.

6.   Allow each sample to completely drain before going to step 7

7.   Add DI water to each funnel and record the time required for another 15 ml of water to flow through each of the treated soils.

8.   Plot on regular graph paper or by computer, with time on the "Y" axis and "Treatment" on the "X" axis, a bar graph showing differences in percolation rates for both increments of water collected for each treatment.

9.   Read Section 9.2 in the text, and explain the differences observed among the treatments.

Comparison Data for Water Percolation Rates in Salt Effected Soils

First Treatment	Time to Collect 15 mls
a. DI water	_____
b. 1N $CaCl_2$	_____
c. 1N NaCl	_____

Second Treatment	Time to collect 15 mls
aa. DI water - DI water	_____
bb. 1N $CaCl_2$ - DI water	_____
cc. 1N NaCl - DI water	_____

10.   Describe any difference you observe in the appearance of the soils in each funnel after collecting the second 15 ml increment of water.

11. Compare the effect calcium and sodium have on the flow rate of water through the soil.

12. Measure the top diameter of the funnel and determine the **Percolation   Rate** in cm/hr for each treatment. Note 15 cm^3 of filtrate/funnel area (cm^2) equals cm water entering the soil per time. Use the time required to collect the 15 mls of solution and convert to cm/hr).

13. Complete the following chart. Show calculations below the chart..

    Influence of Cation Saturation on Water Flow Rates in Soil

Sample Treatment	Flow Rate
aa. DI water - DI water	_____ cm/hr
bb. CaCl$_2$ - DI water	_____ cm/hr
cc. NaCl - DI water	_____ cm/hr

14. Show calculations here.

15. How might a soil that is known to be sodic be reclaimed for food production? (See section 9.5 in the text)

16. At 5 dS/m why is there a difference in the yield reduction of sweet corn and corn for forage? (See Fig. 9.3 and 9.4 in the text, PP 288-299)

Laboratory Exercise XI
Salt Affected Soils

Instructor Copy

Name _____

Section _____

2.   Measure the Electrical Conductivity (EC) of the following liquid samples and
       determine the TDS and Osmotic Potential,(OP) for each sample.
       (The instructor will demonstrate how to use the Conductivity Meter. To obtain an
       exact reading, the EC meter values needs to be adjusted for solution temperature
       and cell constant). Warm water conducts current better than cold water and EC
       cell design affects readings. (See appendix D)

Sample	Adjusted Readings dS/m	TDS, mg/L	OP, kPa
a. Tap water	_____	_____	_____
b. Deionized or distilled water _____		_____	_____
c. Sample from home	_____	_____	_____
d. Soda water	_____	_____	_____

5.   Determine the EC, TDS, and OP for each soil. Since a 2:1 water to soil ratio was
       used, the EC value needs to be doubled to account for a dilution effect.

Sample	Corrected EC dS/m	TDS, mg/L	OP,kPa
Soil A	_____	_____	_____
Soil B	_____	_____	_____

Answer the following questions.

1.   Find the quantity of salt in pounds added to an acre of land by the application of
       an acre-foot of irrigation waters a, b. and c. above. (An acre-foot of water is
       equal to 1.2×10^2 liters of water. Show calculations.

2.   For the soda water, find the amount of TDS in grams one ingests when
       consuming each 20 ounce soft drink.

_____ g

183

Instructor Copy

Name _____

Section _____

3.   Compare the soda water TDS amount with regular tap water TDS and explain
      which is better for quenching thirst. Explain.
      (Note, potable drinking water must contain less than 500 mg/L TDS).

                                    Tap water   _____mg/L

                                    Soda water   _____ mg/L

9.   Read Section 9.2 in the text, and explain the differences observed among the
      treatments.

        Comparison Data for Water Percolation Rates in Salt Effected Soils

            First Treatment   Time to Collect 15 mls
            a. DI water   _____
            b. 1N CaCl₂   _____
            c. 1N NaCl   _____

            Second Treatment   Time to collect 15 mls
            aa. DI water - DI water _____
            bb. 1N CaCl₂ - DI water _____
            cc. 1N NaCl - DI water   _____

10.   Describe any difference you observe in the appearance of the soils in each funnel
      after collecting the second 15 ml increment of water.

11.   Compare the effect calcium and sodium have on the flow rate of water
      through the soil?

Instructor Copy

Name _____

Section _____

12. Measure the diameter of the funnel and determine the **Percolation Rate** in cm/hr for each treatment. Note 15 cm³ of filtrate/funnel area (cm²) equals cm water entering the soil per time. Use the time required to collect the 15 mls of solution and convert to cm/hr).

13. Complete the following chart. Show calculations below the table.

Influence of Cation Saturation on Water Flow Rates in Soil

Sample Treatment	Flow Rate
aa. DI water - DI water	_____ cm/hr
bb. $CaCl_2$ - DI water	_____ cm/hr
cc. NaCl - DI water	_____ cm/hr

14. Show calculations here.

15. How might a soil that is known to be sodic be reclaimed for food production? (See section 9.5 in the text)

16. See Fig. 9.3 and 9.4 in the text, PP 288-299 and explain why, at 5 dS/m, there is a difference in the yield reduction of sweet corn and corn for forage?

187

Appendix A

Equipment and Reagent List

Each student will need four beakers, four flasks, four test tubes, a stirring rod, three funnels, a test tube holder, funnel rack and stand, and a small spatula.

1. Introductory Laboratory

    Three pH kits (Poly-D pH Universal indicator available from the Cal Poly Soil Science Department, three Clinometers, two Munsell Color Books, Table of Optimum pH ranges for plants (See Spurway, C.H. 1941 Michigan Agr.Exp.Sta.Bul 306) two monoliths, two one meter tapes, two 100 ml graduated cylinders for each student, a carton of quartz chips that will fit into the 100 ml graduated cylinders, and lab balances.
    Three samples of soils for pH and color determinations.

2. Minerals, Rocks and Weathering

    Trays of common minerals and rocks (44 specimens from Wards), tray of gleyed soil, Jars containing oxidized and reduced soil, tray of sterilized pipettes, three large mortar and pestle sets. DI water, pH reagent, carton of granitic sand, six trays showing the transition of two parent materials to soil, rock > parent material > soil. Two binocular microscopes and samples of beach sand and crushed sand.

3. Soil Texture, Structure and Water Relations

    Large texture triangle, five known texture samples (sandy loam, loam, clay loam, silt loam, clay), three unknown samples, three percolation tubes, 100 ml graduated cylinders, cheese cloth squares, timer, moisture cans, and a drying oven. Hydrometer, Bouyoucos cylinders, thermometer (optional) four mortar and pestle sets.

4. Bulk Density, Particle Density, and Pore Space

    Six 7.5 cm X 7.5 cm aluminum cores, 100 ml graduated cylinders, sandy loam soil and a granular silt loam soil, 10 - 25 ml graduated cylinders, clean fine sand.

    Organic Matter Decomposition Set-up

    Incubation bottles - three each per student, glass vials, plastic bottle containing 1.0 N NaOH (Caution), tray of moist soil free of debris sieved to four mm, Sample of high C/N organic mater finely ground (sawdust), sample of low C/N organic matter finely ground (alfalfa).

5. Field Laboratory

    Two Field Kits:

    Large nails to mark horizons, two 500 ml water bottles, meter tape, pH kits, dropper bottle with 1.0N HCL, clinometer, Munsell color book, sharpshooter tile spade, two excavated soil pits, knife or spatula, and access to the internet.

7. Organic Matter Lab

    Binocular microscopes, 1.0 N $BaCl_2$ in dropper bottle, $BaCl_2$ waste bottle, carboy of 1.0N ammonium acetate, a 25 ml dispenser (tilt pipette), masking tape, Watman's filter paper #1, 12.5 cm, tray for returned vials, nitrate reagents A & B,(Reagent A very fine dry reducing agent, (a mixture of 300 mesh powdered Zn metal, $MnSO_4$, $BaSO_4$). (Reagent B, fill a 500 ml amber bottle half full with DI water. Add 0.75 g of N-1-Napthy-ethylenediamine Dihydrochloride and mix. Triterate in a small clean acid washed mortar and pestle 0.5 g Of Sulfanilic acid. Transfer to the amber bottle. Add 2.5 mls of conc HCl and bring to 500 mls with DI water. pH of the solution should be 1.5. Store in the dark. Methods of Soil Analysis, Part 3, Ch. 38, PP 1150, SSSA Book Series 1996, Ed. D.L.Sparks.

    Sodium hexametaphosphate solution, (Dissolve 44.6 grams of $Na_4P_2O_7$ * $10H_2O$ in 500 mls of boiling DI water. Add 0.40 g of NaOH and bring to one liter volume), 5 - 50 ml graduated cylinders.

8. Soil Testing and Fertilizers

    Whatman filter paper #1 12.5 cm, plastic funnels, 50 ml tilt pipette, 10 ml graduated cylinders, fertilizer samples commercial, organic, specialty, fertilizer bags, DAP fertilizer, 4 bottles containing 1.0 N NaOH, 4 dropper bottles containing saturated ammonium (add 15 g of ammonium oxalate to 400 mls of hot DI water, dissolve and bring to 500 mls volume), colorimeter or nepholometer, standard Ca curve.

9. Soil Survey Reports

    Large laminated Soil Map, Modern Soil Survey Reports, clear plastic sheet, erasable markers, rulers, stereoscope scope/s and aerial map pairs.

11. Salt Affected Soil

    Whatman filter paper # 1, three plastic funnels per student, clay loam soil, 1.0 N NaCl, 1.0 N $CaCl_2$, 0.01 N KCL used to determine the EC meter cell constant, deionized water, a solu-bridge that measures either ohms, mmhos/cm, or μmhos/cm .

Land Capability Classification
Guide

The Land Capability Classification of soils is a grouping of individual soils into classes based upon the influence climate has on soil use, management, and productivity.

There are two general divisions, each containing four classes: (1) land suited for cultivation and other intensive uses, and (2) land limited in use and not suitable for cultivation. In division (1) are classes I, II, III, IV and in division (2) are classes V, VI, VII, VIII.

Each class is further subdivided into subclasses to indicate the major limitation of the class for use or its conservation problem. The subclasses are:

> "e" erosion hazard
> "w" wetness problems (poorly drained, floods)
> "s" soil problem (shallow, stony, salinity etc ..)
> "c" climate (extreme temperatures, arid)

LAND SUITED FOR CULTIVATION AND OTHER INTENSIVE USES

CLASS I

Soils in Class I have few or no limitations or hazards for intensive use. They can be used safely for cultivated crops, pasture, range, and woodland.

CLASS II

Soils in Class II require thoughtful soil management to prevent deterioration. Limitations include gentle slope, less than 36 inches deep, unfavorable structure, slight salinity, ocassional overflow, permanent wetness correctable by drainage. These soils may be used for most cultivated crops, pasture, range, and woodland.

CLASS III

Soils in Class III have severe limitations and require special conservation practices. Moderately steep slopes, high erosion potential, frequent overflow, slow permeability of subsoil, shallow depth to bedrock, root restricting pans,

moderate salinity, and low moisture holding capacity.

CLASS IV

Soils in Class IV have very severe limitations. Factors restricting their use include steep slopes, very high erosion potential, shallowness, excessive wetness, severe salinity, or low water holding capacity. These soils may be used for a few crops, pasture, woods.

LAND LIMITED IN USE GENERALLY NOT SUITED FOR CULTIVATION

CLASS V

These soils are level and have little or no erosion hazard but are impractical to use for intensive croping. Limitations include frequent flooding, excessive stones, ponding. Suitable for pasture, range, or woodland.

CLASS VI

These soils have severe limitations for use without costly reclaimation. Problems include steep slopes, stoniness, severe erosion potential, shallow root zone, salinity, and low moisture capacity. These soils are suited only for pasture, range, and woodland. Some can be used for orchards, groves, and tree farms with intensive management.

CLASS VII

Soils in this class have very severe limitations that restrict their use to grazing, woodland, and rangeland. Problems include very steep slopes, shallow soil, excessive stones, a severe erosion hazard that can not be corrected.

CLASS VIII

Soils and landforms that have limitations that restricts their use to recreation, wildlife, and watersheds. Included in this class are beaches, badlands, rock outcrops, river wash, mine tailings, and other nearly barren lands.

Source: Agriculture Handbook No. 210

Appendix C

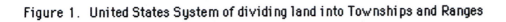

Figure 1. United States System of dividing land into Townships and Ranges

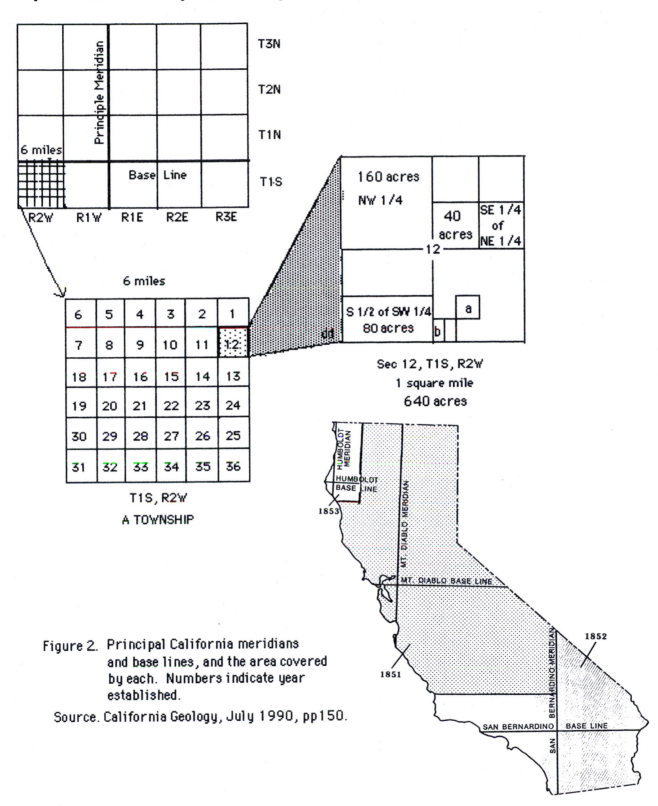

Figure 2. Principal California meridians
and base lines, and the area covered
by each. Numbers indicate year
established.

Source. California Geology, July 1990, pp150.

Appendix D

Chart for correcting Electrical Conductivity and Resistance readings on water and soil extracts to the standard temperature of 25°C. Chart taken from USDA Agriculture Handbook 60 Page 90, 1954.

$$EC_{25} = EC_t \times f_t; \quad EC_{25} = (k/R_t) \times f_t; \quad R_{25} = R_t/f_t$$

°C.	°F.	f_t	°C.	°F.	f_t	°C.	°F.	f_t
3.0	37.4	1.709	22.0	71.6	1.064	29.0	84.2	0.925
4.0	39.2	1.660	22.2	72.0	1.060	29.2	84.6	.921
5.0	41.0	1.613	22.4	72.3	1.055	29.4	84.9	.918
6.0	42.8	1.569	22.6	72.7	1.051	29.6	85.3	.914
7.0	44.6	1.528	22.8	73.0	1.047	29.8	85.6	.911
8.0	46.4	1.488	23.0	73.4	1.043	30.0	86.0	.907
9.0	48.2	1.448	23.2	73.8	1.038	30.2	86.4	.904
10.0	50.0	1.411	23.4	74.1	1.034	30.4	86.7	.901
11.0	51.8	1.375	23.6	74.5	1.029	30.6	87.1	.897
12.0	53.6	1.341	23.8	74.8	1.025	30.8	87.4	.894
13.0	55.4	1.309	24.0	75.2	1.020	31.0	87.8	.890
14.0	57.2	1.277	24.2	75.6	1.016	31.2	88.2	.887
15.0	59.0	1.247	24.4	75.9	1.012	31.4	88.5	.884
16.0	60.8	1.218	24.6	76.3	1.008	31.6	88.9	.880
17.0	62.6	1.189	24.8	76.6	1.004	31.8	89.2	.877
18.0	64.4	1.163	25.0	77.0	1.000	32.0	89.6	.873
18.2	64.8	1.157	25.2	77.4	.996	32.2	90.0	.870
18.4	65.1	1.152	25.4	77.7	.992	32.4	90.3	.867
18.6	65.5	1.147	25.6	78.1	.988	32.6	90.7	.864
18.8	65.8	1.142	25.8	78.5	.983	32.8	91.0	.861
19.0	66.2	1.136	26.0	78.8	.979	33.0	91.4	.858
19.2	66.6	1.131	26.2	79.2	.975	34.0	93.2	.843
19.4	66.9	1.127	26.4	79.5	.971	35.0	95.0	.829
19.6	67.3	1.122	26.6	79.9	.967	36.0	96.8	.815
19.8	67.6	1.117	26.8	80.2	.964	37.0	98.6	.801
20.0	68.0	1.112	27.0	80.6	.960	38.0	100.2	.788
20.2	68.4	1.107	27.2	81.0	.956	39.0	102.2	.775
20.4	68.7	1.102	27.4	81.3	.953	40.0	104.0	.763
20.6	69.1	1.097	27.6	81.7	.950	41.0	105.8	.750
20.8	69.4	1.092	27.8	82.0	.947	42.0	107.6	.739
21.0	69.8	1.087	28.0	82.4	.943	43.0	109.4	.727
21.2	70.2	1.082	28.2	82.8	.940	44.0	111.2	.716
21.4	70.5	1.078	28.4	83.1	.936	45.0	113.0	.705
21.6	70.9	1.073	28.6	83.5	.932	46.0	114.8	.694
21.8	71.2	1.068	28.8	83.8	.929	47.0	116.6	.683

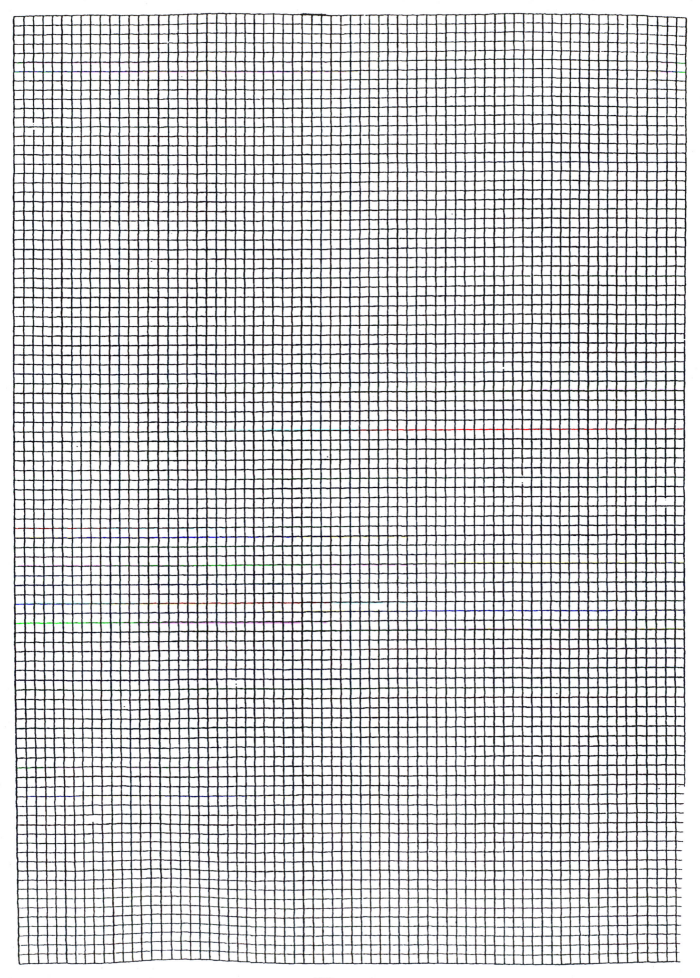